Wiesława Widłak

Molecular Biology

Not Only for Bioinformaticians

 Springer

Author

Wiesława Widłak
Maria Skłodowska-Curie Memorial Cancer Center
and Institute of Oncology, Gliwice Branch
Wybrzeże Armii Krajowej 15, 44-101 Gliwice, Poland
E-mail: wwidlak@io.gliwice.pl

ISSN 0302-9743   e-ISSN 1611-3349
ISBN 978-3-642-45360-1   e-ISBN 978-3-642-45361-8
DOI 10.1007/978-3-642-45361-8
Springer Heidelberg New York Dordrecht London

Library of Congress Control Number: 2013955613

CR Subject Classification (1998): J.3

LNCS Sublibrary: SL 8 – Bioinformatics

Typesetting: Camera-ready by author, data conversion by Scientific Publishing Services, Chennai, India

Printed on acid-free paper

Springer is part of Springer Science+Business Media (www.springer.com)

Preface

The book is addressed to beginners in molecular biology, especially to computer scientists who would like to work as bioinformaticians (computational biologists). Bioinformatics, which could be defined as the application of computer science and information technology to the field of biology and medicine, has been rapidly developing over the past few decades. It generates new knowledge as well as the computational tools to create that knowledge. Certainly, understanding of the basic processes in living organisms is indispensable for bioinformaticians in the creation of knowledge. The book contains information about these processes presented in a condensed manner. Additionally, principles of several high-throughput technologies in molecular biology, which need the assistance of bioinformaticians, are explained from a biological point of view. Although some processes observed in living organisms are described here in a more detailed fashion, the book may be treated as an introduction to molecular biology. Thus it could be the first choice for people who are unfamiliar with the subject.

Contents

Chapter 1
Cells and Viruses

Cells are the smallest structural component of all known living organisms capable of self-maintenance and reproduction. Although cells vary greatly in their appearance or size, their structure is basically similar. Even the plant and animal cells show a significant degree of similarity in their overall organization.

There are two types of cells: **eukaryotic** and **prokaryotic**. The main difference between them is the method of genetic material storage: in eukaryotic cells — in an isolated nucleus, in prokaryotic cells — directly in the cytoplasm (there is no nucleus). Prokaryotic cells are usually independent (unicellular), while eukaryotic cells are often found in multicellular organisms.

Eukaryotic organisms are organisms composed of eukaryotic cells. We classify here: animals, plants, fungi, and protists (including organisms that do not fit into the remaining groups). A new classification of prokaryotic organisms distinguishes archaea (formerly known as archaebacteria) and bacteria (formerly eubacteria). Both are single cell organisms. Archaea live in extreme environments. They are divided into three main groups on the basis of living environment: **extreme thermophiles** (live in high-temperature environments, rich in sulfur compounds and low pH, such as hot springs and geysers), **extreme halophiles** (live in environments with high concentrations of NaCl, such as the Dead Sea), and **methanogens** (live in anaerobic environments such as swamps, produce methane and do not tolerate oxygen).

1.1 Cell Structure

All cells contain cytoplasm, cell membrane, and DNA (the carrier of genetic information). The cytoplasm is the liquid filling of the cell, with high density, holding all the cell's internal sub-structures (called the organelles), except for the nucleus. It contains also insoluble substances (so called cytoplasmic inclusions) and free ribosomes. A cell membrane surrounds the cytoplasm and isolates cell from the environment. The cell membrane is composed of the phospholipid bilayer, which contains proteins and carbohydrates. The phospholipid bilayer is also the basic component of other biological membranes. DNA (deoxyribonucleic acid) is the genetic material. Eukaryotic cells possess several pieces of DNA, located in organelles surrounded by a phospholipid membrane (in the nucleus and mitochondria). Prokaryotic cells possess a single DNA molecule, directly in the cytoplasm.

1.1.1 Prokaryotic Cell Structure

A prokaryotic cell is simpler and smaller than an eukaryotic cell. A typical bacterial cell is $5\mu m$ long and $1\mu m$ wide. There are also smaller bacteria ($0.5\mu m$ in diameter), or bigger. A prokaryotic cell contains DNA, cytoplasm, plasma membrane, and also some of the following components: cell wall, capsule, slime, flagellum and fimbriae/pili (Fig 1.1). It does not contain organelles. Prokaryotes are always single-cell organisms; however most are capable of forming stable aggregate communities. When such communities are encased in a stabilizing polymer matrix ("slime"), they may be called "biofilms". Bacterial biofilms may be 100 times more resistant to antibiotics than free-living unicells and may be nearly impossible to remove from surfaces once they have colonized them.

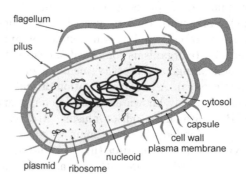

Fig. 1.1 Generalized structure of the prokaryotic cell

A bacterial cell wall contains a unique structure called peptidoglycan. Archaea do not have peptidoglycan, which means that they are resistant to many antibiotics interacting with the cell wall. Due to differences in cell wall structure and the effect of staining by the Gram dye, bacteria can be divided into two groups: Gram-negative and Gram-positive. The cell wall of Gram-negative bacteria consists of three layers: the periplasmic space, which is an open area on the outside of the plasma membrane, then a thin layer of peptidoglycan, and an outer membrane surrounding the peptidoglycan.

In the cell wall of Gram-positive bacteria, there is no periplasmic space and outer membrane, while the peptidoglycan layer is thicker. As a result, they are more sensitive to the lysozyme (an enzyme that is naturally present e.g. in saliva or tears and is able to hydrolyze bacteria cell wall), and penicillin (antibiotic derived from fungi).

Capsule and slime form hydrophilic surrounding of the cell wall in most bacteria. Capsule sticks more closely to the wall than slime. Flagellum is a tail-like projection that protrudes from the cell body. It facilitates the movement of mobile bacteria. Fimbriae and pili are structures similar to short hairs. They attach cells to other cells, which is important when infecting other organisms.

Spores are small, haploid and unicellular structures used for reproduction. They can be found in plants, algae, fungi, protozoa and bacteria. Bacteria spores have thick wall, which allows them to survive in wide range of temperatures, humidity and other unfavorable conditions, which kills vegetative forms exhibiting normal life activity.

All information essential for the bacteria survival (in the environment typical for it) are saved in the chromosome (bacterial nucleoid). Bacterial chromosome is usually large, circular piece of DNA, but sometimes linear chromosomes and several copies of them occur. Bacteria may have also plasmids. These are (usually) circular DNA molecules that are replicated independently of the chromosome. Plasmids often contain genes that could be very useful, for example include information how to use a compound present in environment. If this compound is in the environment, the bacteria gain an obvious advantage over other bacteria, which do not have the ability to decompose that compound. Plasmids can be different in sizes, from tiny to megaplasmids exceeding the size of the genomes of some bacteria. Plasmids are natural vectors, by which continuous exchange of genetic information among bacteria (sometimes even completely unrelated) occurs. For example, many antibiotic resistance genes reside on plasmids, facilitating their transfer. Plasmids serve as important tools in bioengineering, where they are commonly used to multiply or express particular genes.

1.1.2 Eukaryotic Cell Structure

Eukaryotic cells contain organelles, which are defined as membrane-enclosed structures (such as nucleus, mitochondria, chloroplasts, endoplasmic reticulum, Golgi apparatus, lysosomes, vacuoles or peroxisomes). Animal cell is surrounded only by the plasma membrane, while the plant cell has an additional layer called the cell wall, which is formed from cellulose and other polymers (Fig. 1.2). Typical size of eukaryotic cells varies from 5 to 50 μm.

The nucleus is the largest structure in the eukaryotic cell. It is not part of the cytoplasm, which can be defined as everything that is surrounded by a

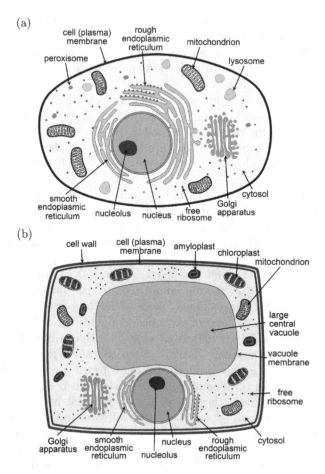

Fig. 1.2 A typical eukaryotic cell illustration: (a) animal; (b) plant

cell membrane, excluding the nucleus. The term "cytoplasm" is however often used in the narrower sense, meaning only the cytosol. **Cytosol** is the cytoplasm after exclusion of all organelles. It is a liquid, water colloid containing proteins (suspended or dissolved), lipids, fatty acids, free amino acids and minerals (e.g. calcium, magnesium, sodium). Complex network of protein fibrils (microfilaments) and microtubules, forming the **cytoskeleton** (which allows the cell to maintain its shape and size), is also an important component of the cytoplasm.

1.2 Biological Membranes and Their Lipid Component

All biomembranes (cell membrane and membranes of all intracellular organelles) are highly selective permeable barriers, setting out the boundaries of cells and organelles. They are built from lipids and proteins. The lipids existing in biological membranes there are mainly phospholipids and cholesterol. Carbohydrates may be attached to proteins and phospholipids present in the membranes — forming respectively glycoproteins and glycolipids (Fig. 1.3). The membranes are liquid layered structures: the majority of lipid and proteins molecules can move fast along the membranes. It is worth noting that the lipids, besides cell membrane building function, can also be used as "fuel particles", energy storing molecules and as signaling molecules (ch. 8).

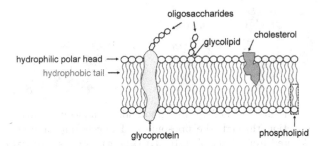

Fig. 1.3 Scheme of a typical biological membrane (lipid-protein bilayer)

1.2.1 Phospholipids

Phospholipid molecule is composed of hydrophilic (with high affinity for water) polar group, so-called "head", and hydrophobic (with low affinity for water) "tail". Hydrophobic tail is made up of two fatty acid chains (Fig. 1.4). Most polar phospholipids are phosphoglycerides that contain glycerol connecting the head with the tail. For example phosphatidylcholine, phosphatidylserine, phosphatidylethanolamine or phosphatidylinositol belong to this group. Chains of fatty acids in biological membranes usually contain an even number of carbon atoms, generally from 14 to 24. There are saturated (the neighboring carbon atoms are linked by a single bond) or unsaturated fatty

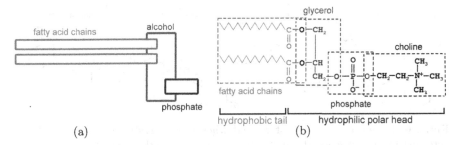

Fig. 1.4 The chemical structure of phospholipids: (a) general model, (b) structure of phosphatidylcholine

acids (when some of the neighboring carbon atoms are connected by a double bond). The chain length and degree of saturation of fatty acids in lipids has significant influence on the fluidity of biological membranes.

When phospholipid molecules are placed in water, the hydrophilic heads are oriented facing the water and the hydrophobic tails avoid water. As result, the structure called the bilayer sheet, liposome, or micelle may be formed (Fig. 1.5).

Fig. 1.5 The structures formed by phospholipids in aqueous solutions. The figure presents a section through the liposome and micelle and a fragment of the bilayer sheet. Altered from: http://upload.wikimedia.org/wikipedia/commons/c/c6/Phospholipids_aqueous_solution_structures.svg

1.2.2 Cholesterol and Steroids

Cholesterol does not exist in most prokaryotic cells, whereas it is a common component of mammalian cell membranes. It fulfills a critical role in the regulation of membrane fluidity. Also synthesis of steroid hormones is based on

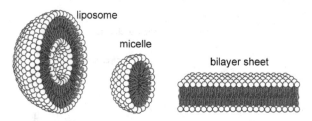

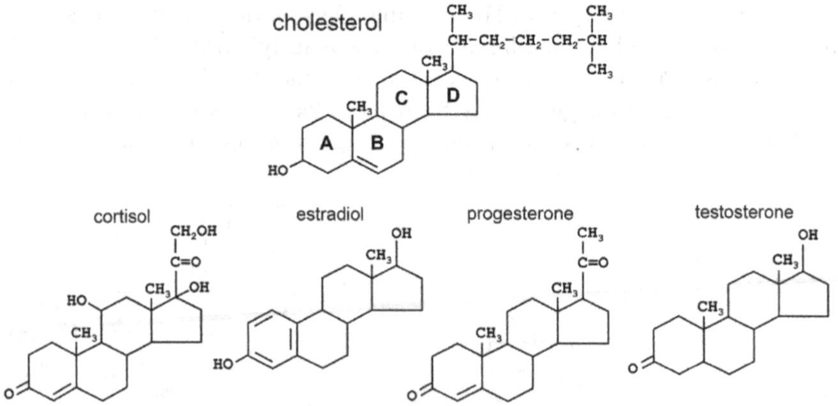

Fig. 1.6 The structure of cholesterol and other important steroids. They are characterized by the presence of four hydrocarbon rings, marked here as A, B, C and D. Due to the presence of the hydroxyl group (-OH) cholesterol has amphipathic properties (it has hydrophobic and hydrophilic parts).

cholesterol (Fig. 1.6; ch. 8.1). Accumulation of cholesterol in an artery wall plays a key role in atherosclerosis — disorder, which can lead to heart attack or stroke.

1.3 Nucleus

The nucleus consists of a nuclear envelope (constructed from two membranes; Fig. 1.7), the nucleolus and nucleoplasm. The outer nuclear membrane is continuous with the membrane of the rough endoplasmic reticulum (RER). The space between the membranes (perinuclear space) is continuous with the RER lumen. With the inner face of the nuclear envelope the nuclear lamina is associated, which is built from fibrous proteins called lamins. The nuclear lamina binds inner nuclear membrane with genetic material. The genetic material formed in the chromatin/chromosomes is located mainly in nucleoplasm. However, the parts of various chromosomes, which contain groups of genes encoding rRNA (ch. 3.3.3), could be concentrated in one region forming the nucleolus. Its main role is the production of rRNA.

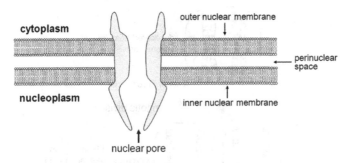

Fig. 1.7 Scheme of a nuclear envelope which is built from two layers of double lipid membrane. In the nucleus of mammals there are about 4,000 nuclear canals (pores) present, each composed of more than 100 various proteins.

1.3.1 Chromosomes and Karyotype

Chromosome is condensed structure, which is formed from a single DNA molecule and associated proteins. In cells which are not dividing at the moment, chromosomes are not visible in light microscope, because DNA together with accompanying proteins (i.e. chromatin) is loosely spreaded across the nucleus. During cell division, in the stage of metaphase (see ch. 7), chromatin condenses and becomes visible in the form of chromosomes.

Each chromosome has the short and long arm (designated as p, from the "petit" and q, from the "queue"). The arms are separated by narrow region called the centromere (Fig. 1.8). The chromosomes of the same shape and size, containing similar genetic information (i.e. genes) are called homologous chromosomes. In diploid cells, one of them comes from the mother and the other from the father. During cell division, at metaphase, replicated chromosomes contain two sister chromatids linked with centromere. Sister chromatids are identical copies of the chromosome: they contain the same genes and the same alleles (i.e. versions of genes in a particular location — locus — on the homologous chromosome). However, homologous chromosomes contain the same genes, but alleles can vary. During cell division pairs of chromosomes are separated into two daughter cells.

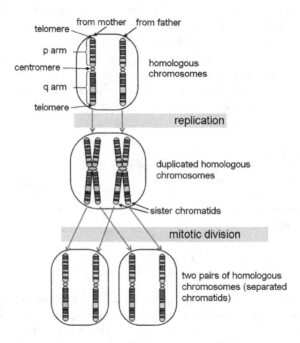

Fig. 1.8 Homologous chromosomes during mitotic cell division

Chromosomes are visible under the microscope only after staining. Then, they look like strings with dark and light transverse stripes. A set of chromosomes present in the cell during metaphase is called a **karyotype**. Human germ cells (egg or sperm) contain one set of chromosomes (haploid cell, 1n), that is 23 pieces: 22 autosomes and the X or Y. Somatic cells (remaining cells of the body) normally contain two homologous sets of chromosomes, i.e. 46 chromosomes (diploid, 2n). One set comes from mother, the other — from father. In other species the number of chromosomes varies from 1 to 1260.

Human somatic cells contain two chromosomes that determine sex: XX in females and XY in males. In germ cells there is only one sex chromosome: in female it is always X chromosome, in male — X or Y.

1.4 Cytoplasmic Organelles

Organelles can be defined as intracellular, bounded by membrane, protein-lipid structures (the cell nucleus is also an organellum). In cytoplasm there are present mitochondria, chloroplasts, endoplasmic reticulum, Golgi apparatus, peroxisomes, lysosomes, vacuoles, and glyoxysomes.

1.4.1 Mitochondria

Eukaryotic cells have a lot of mitochondria — they can be up to a quarter of the cytoplasm volume. They are elongated organelles with a diameter ranging from 0.5 to 10 μm (the size corresponds to a bacterial cell). Mitochondria are composed from two membranes: outer and inner, which is strongly folded. Mitochondria have their own DNA (defined as the mtDNA) coding for proteins and RNA. However the majority of mitochondrial proteins are encoded by nuclear DNA.

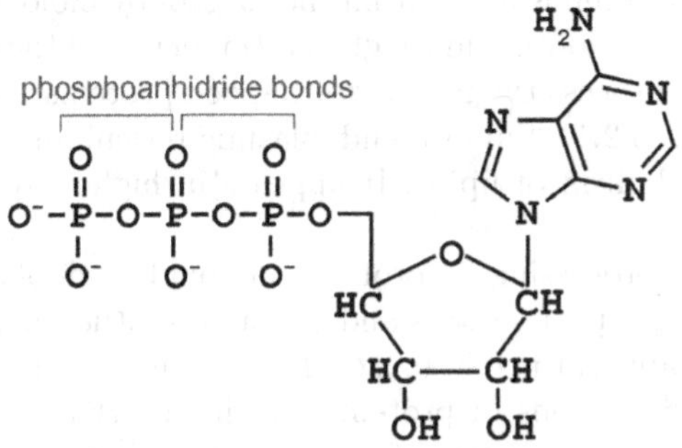

Fig. 1.9 Chemical structure of ATP (adenosine triphosphate)

The main task of mitochondria is the production of ATP (adenosine triphosphate), which provides source of energy for most cellular processes (Fig. 1.9). Energy is stored in phosphate bonds and can be released during the hydrolysis of ATP to ADP (adenosine diphosphate) and AMP (adenosine monophosphate).

In animal cells, the catabolism (degradation of glucose and fatty acids) is the main source of energy for ATP synthesis. The production of ATP takes place in the inner mitochondrial membrane. Energy is transferred from electron transport chain to the ATP synthase by movements of protons across this membrane. As a result of disturbances in electron transport, the free radicals (and other reactive oxygen species) can be generated, which can lead to damage of cell structures and to cell senescence.

1.4.2   Chloroplasts

Cloroplasts, like mitochondria, are composed from inner and outer membrane. Chloroplasts also have their own DNA, but most of their proteins are encoded by nuclear DNA. The inner membrane of the chloroplast forms thylakoids, that contain chlorophyll. It absorbs light necessary for the light reaction of photosynthesis. In this stage, thanks to the energy absorbed from light quanta, oxygen molecule and hydrogen ions are formed from two water molecules. This results in forming of the concentration gradient in the thylakoid membrane. The flow of ions through the membrane is connected with the synthesis of ATP. Stroma is the liquid component of the chloroplast, in which, using the energy from ATP, CO_2 is assimilated and synthesis of saccharides occurs (dark reactions of photosynthesis).

1.4.3   Endoplasmic Reticulum and Golgi Complex

Endoplasmic reticulum (ER) forms an irregular network of cisternae, vesicles and tubules inside the cell isolated from the cytosol by biological membranes. It exists in two forms: smooth and rough. The main role of the rough endoplasmic reticulum is to process the newly synthesized proteins, so it is associated with ribosomes (ch. 6.2.1). Smooth endoplasmic reticulum is involved in the synthesis and metabolism of lipids. It appears in highest quantities in liver cells (hepatocytes).

After the initial processing of proteins in rough endoplasmic reticulum, they are closed in transport vesicles and directed to other membranous structures — the Golgi apparatus (ch. 6.2.2). These structures are the main place of sorting and modifications of proteins and lipids. Here, glycoproteins are produced and they are transported further to their destinations. Golgi apparatus is also a place of synthesis of polysaccharides and mucopolysaccharides.

1.4.4   Peroxisomes, Lysosomes, Vacuoles, and Glyoxysomes

Peroxisomes contain enzymes, which allow degradation of amino acids and fatty acids. In these reactions the hydrogen peroxide is formed, which is harmful to cells. It is converted to water and oxygen by an enzyme called catalase.

The main function of lysosomes is degradation of unnecessary macromolecules. Macromolecules are hydrolyzed by the respective enzymes: nucleic acids (DNA and RNA) by **nucleases**, proteins by **proteases**, and lipids by **lipases**. Lysosomes are present only in animal cells. In plant cells their functions are performed by the vacuoles.

Vacuoles store water, ions, saccharides, amino acids and other small molecules. They can also store unnecessary cell products, which can undergo degradation here. Vacuoles usually fill about 30% of the cell volume, but can increase its volume up to 90%.

Glyoxysomes are found mainly in plant seeds. Their main function is to convert fatty acids into acetyl-CoA molecules, which in glyoxylate cycle are converted into monosaccharides. This mechanism is important in the use of fat reserves of oilseeds. It occurs during germination and when it is necessary to launch a non-carbohydrate energy reserves in plants. Glyoxysomes and peroxisomes are also called microbodies.

1.5 Viruses, Bacteriophages, and Virophages

Viruses are macromolecular complexes built from proteins and nucleic acids. Viruses are not able to replicate by themselves — so they can not be treated as living organisms. Replication of viruses is possible only within other cells, using enzymes from infected cells. Because of that viruses are intracellular parasites. Most viruses have a diameter between 10 and 300 nm (but discovered in 2013 Pandoraviruses have a size approaching $1\mu m$). The molecule of the virus (**virion**) consists of one or more molecules of DNA or RNA (as a carrier of genetic information of the virus) and a protein shell called a **capsid**. Some viruses have an additional lipid envelope which surrounds the capsid. Most viruses are pathogens — after penetration into living cells they cause the diseases. Various viruses specifically attack different types of cells (e.g. the HIV virus attacks only the immune cells, mainly T lymphocytes). When a virus encounters "his" cell, it attaches to the cell membrane using specific receptors (proteins on the surface of host cells, which normally have other functions). Then, it can inject its genetic material into the cell or whole virus is absorbed entirely by the cell. Inside the cell, based on the information contained in the genetic material of the virus, there are created new virus molecules. They can remain in the cell for a long latent phase, or leave after the cell destruction. It sometimes happens, that the virus remains in the cell in form of genetic material integrated with the host genome.

Viruses can also infect bacteria. Such viruses are called bacteriophages (or commonly phages). Bacteriophages can be used for treating bacterial infections as an alternative to antibiotics. In biotechnology and genetic engineering, phages and other viruses are used to transfer genetic information between cells (as so-called "vectors").

There are also viruses that infect other viruses. The first known, called Sputnik, was discovered in 2008. It is a subviral agent that reproduces in amoeba cells which are already infected by a certain helper virus.

Sputnik uses the helper virus's machinery for reproduction and inhibits replication of the helper virus. Viruses that depend on co-infection of the host cell by helper viruses are known as satellite viruses. They act as a parasite of helper viruses. In analogy to the term bacteriophage they were called virophages.

The risk associated with viral infections may be very different. Many viral infections do not cause diseases, for example, most people are infected with cytomegalovirus (CMV), however, do not show symptoms of the disease. There are also viruses that often lead to death of the organism, such as coronavirus which cause Severe Acute Respiratory Syndrome (SARS) or the HIV virus causing AIDS.

1.5.1 Classification of Viruses

The classification of viruses could be based on the type of genetic material held by the virus and on strategy of genetic material replication (the Baltimore classification). Besides double-stranded DNA, the carrier of genetic information in prokaryotic and eukaryotic cells, also single-stranded DNA or RNA can be a genetic material in viruses. Viruses were divided into seven classes (ds — double strand, ss — single strand):

     I. dsDNA viruses — contain double-stranded DNA;
    II. ssDNA viruses — contain single-stranded DNA;
   III. dsRNA viruses — contain double-stranded RNA;
   IV. ssRNA(+) viruses — contain single-stranded RNA, (+) sense;
    V. ssRNA (-) viruses — contain single-stranded RNA, (-) sense;
   VI. RNA viruses that use virally encoded reverse transcriptase to produce DNA from the initial virion RNA genome;
  VII. DNA viruses that use virally encoded reverse transcriptase.

The last two classes of viruses are known as retroviruses. Rewriting of the genetic information from RNA to DNA is an essential part of their replication cycle. This reaction is catalyzed by reverse transcriptase — an enzyme encoded by the virus genome (ch. 5.9).

References

1. Adams, M. (ed.): Subcellular Compartments. In: Miko, I. (ed.) Cell Biology. Nature Education (2011), http://www.nature.com/scitable/topic/subcellular-compartments-14122679
2. Coté, G., De Tullio, M. (eds.): Cell Origins and Metabolism. In: Miko, I. (ed.) Cell Biology. Nature Education (2011), http://www.nature.com/scitable/topic/cell-origins-and-metabolism-14122694
3. Davidson, M.W.: Introduction to Cell and Virus Structure. In: Molecular Expressions Website. The Florida State University (2000), http://micro.magnet.fsu.edu/cells/index.html

4. La Scola, B., Desnues, C., Pagnier, I., et al.: The virophage as a unique parasite of the giant mimivirus. Nature 455, 100–104 (2008)
5. Liang, B.: Construction of the Cell Membrane. Wisc-Online.com (2001), http://www.wisc-online.com/objects/ViewObject.aspx?ID=AP1101
6. Lodish, H., Berk, A., Zipursky, S.L., et al.: The Dynamic Cell. In: Molecular Cell Biology, 4th edn., W.H. Freeman, New York (2000)
7. Mc Grath, S., van Sinderen, D. (eds.): Bacteriophage: Genetics and Molecular Biology, 1st edn. Caister Academic Press, Norfolk (2007)
8. O'Connor, C.M., Adams, J.U.: Essentials of Cell Biology. NPG Education, Cambridge (2010)
9. Ogata, H., Claverie, J.M.: Microbiology. How to Infect a Mimivirus. Science 321, 1305–1306 (2008)

Chapter 2
Protein Structure and Function

Proteins are linear chains of amino acids and are fundamental components of all living cells (along with carbohydrates, fats and nucleic acids). They make up half the dry weight of an *Escherichia coli* (the most widely studied prokaryotic model organism) cell, whereas other macromolecules such as DNA and RNA make up only 3% and 20%, respectively. Within cells, as well as outside, proteins serve a myriad of functions. The chief characteristic of proteins that allows their diverse set of functions is their ability to bind other molecules (proteins or small-molecule substrates) specifically and tightly.

2.1 Amino Acids

Amino acids serve as the building blocks of proteins. Amino acid (in short: aa) is defined as a molecule containing an amine group (-NH$_2$), carboxyl group (-COOH) and the variable group denoted as R, different among different amino acids (Fig. 2.1). R group is also called the side chain. The overall amino acid formula can be represented as: R-**CH(NH2)-COOH**. An average molecular weight is about 135 daltons.

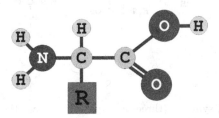

Fig. 2.1 Schematic diagram of the amino acid: amine group — on the left, carboxylic acid group — on the right, R — side chain (variable). The carbon atom located between the amine group and carboxylic acid group is called α-carbon.

In aqueous solutions amino acids can be found in the form of inner salts (they have amphiprotic properties) — they can attach or detach protons to the functional groups (i.e. amine and carboxylic acid groups). Therefore, the isoelectric point (pI) is an important criterion, which characterizes amino acids. pI is at such a pH at which a population of molecules contains approximately the same number of positive as negative charges (the total charge is zero).

There are known about 300 naturally found amino acids, but only 20 of them make up the proteins (Fig. 2.2). They are known as **proteinogenic amino acids**. They belong to the α-amino acids, in which both functional groups and side R chain are combined with one carbon atom, called α-carbon.

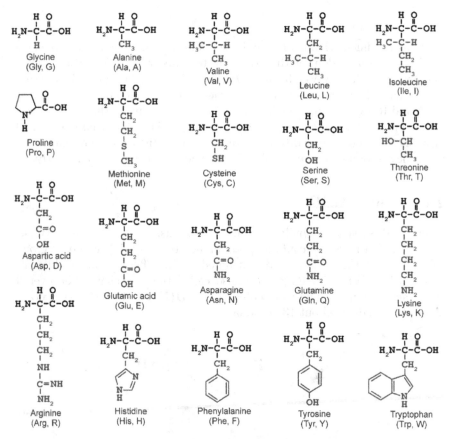

Fig. 2.2 The names, symbols (three- and one-letter), and the chemical structure of proteinogenic amino acids

These 20 are encoded by the universal genetic code (ch. 3.5). Nine standard amino acids are called "essential" for humans because they cannot be created from other compounds by the human body, and so must be taken in as food.

Proteinogenic amino acids (except for glycine, which has no asymmetric carbon atom in the molecule because hydrogen is here the R group), are optically active compounds. Proteinogenic amino acids belong to the group of L-amino acids, which means that they have an amino group on the left side of the R chain.

Based on physical and chemical properties of the side chain, amino acids are divided into:

acidic (aspartic acid and glutamic acid; named also aspartate or glutamate) — in a neutral solution a side chain containing carboxyl group may lose a proton and the amino acid becomes negatively charged;

basic (lysine, arginine, histidine) — in a neutral solution a side chain containing amino or imino group can gain a proton and become positively charged (the interactions between the positively and negatively charged side chains of acidic or alkaline amino acids can lead to the formation of "salt bridges" which stabilize protein structure);

aromatic (tyrosine, tryptophan, phenylalanine) — their side chain contains an aromatic ring;

sulfur (cysteine and methionine) — their side chain contains sulfur (disulfide bonds formed between two cysteines strongly stabilize the structure of the protein, while from methionine the synthesis of all proteins starts);

uncharged hydrophilic (serine, threonine, asparagine, glutamine) — they have a side chain which contains a hydrophilic (but not undergoing ionization) functional group, for example the hydroxyl group (such chains are involved in the formation of hydrogen bonds, stabilizing the protein structure);

inactive hydrophobic (glycine, alanine, valine, leucine, isoleucine) — they have a hydrocarbon side chain without functional groups (such chains can participate in hydrophobic interactions, in protein structure they are usually directed to the inside);

special structure — in this group is proline, in which the R group is directly linked to the amine group (the spatial structure of the polypeptide chain is turned and stiffened in a place of proline).

2.2 Peptides and Proteins

Amino acids can be linked together by peptide bonds in varying sequences to form a vast variety of peptides or proteins. Peptides are distinguished from proteins on the basis of size, typically containing less than 50 monomer units. Oligonucleotides consist of between 2 and 20 amino acids, while polypeptides refer to the longer peptides (the size boundaries are arbitrary).

Proteins consist of one or more polypeptides arranged in a biologically functional way and are often bound to cofactors, or other proteins.

The molecular weight of most polypeptides varies between 5,500 to 220,000. The mass of the proteins is given in Daltons (named after John Dalton, who has formulated atomic theory of matter building): one dalton corresponds to one atomic mass unit. Thus, a protein with a molecular weight of 50,000 has a weight of 50,000 daltons, or 50 kDa (kilodaltons).

Peptide bond is formed by the **condensation reaction**. It is an amide bond formed between the nitrogen from α-amine group of one amino acid and carbon from α-carboxyl group of another amino acid (Fig. 2.3). Peptide bond is a flat and rigid structure, while bonds between α-carbon and amine group or carboxyl group have a possibility to rotate. The line formed by the repeating peptide bonds is the **backbone** of the peptide chain.

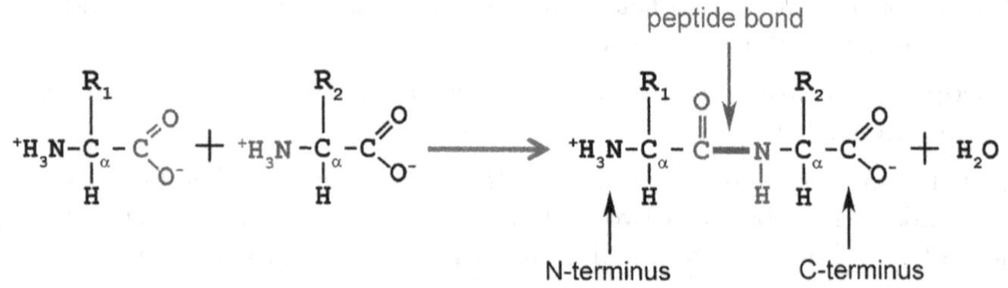

Fig. 2.3 Formation of the peptide bond by condensation reaction. Chain of amino acids linked by peptide bonds forms a peptide or a protein.

2.3 Levels of Protein Structure

Structural features of proteins are usually described at four levels of complexity:

1. Primary structure: the linear arrangement of amino acids (also called a residue) in a protein and the location of covalent linkages such as disulfide bonds between amino acids. It is always given starting from the N-terminus (where free α-amine group is present). The C-terminus it is an opposite end, with a free carboxyl group (Fig. 2.4). The primary structure of a protein can readily be deduced from the nucleotide sequence of the corresponding messenger RNA.
2. Secondary structure: areas of folding or coiling within a protein, which are stabilized by hydrogen bonding. A polypeptide chain is usually arranged in the shape of α-helix or β-strand.
3. Tertiary structure: the final three-dimensional structure of a protein, which results from a large number of non-covalent interactions between amino acids.

NH₃⁺ – $\overset{1}{\text{K}}$ETAAAKFERQHMDSSTSAASSSN$\overset{20}{\text{Y}}$ C
                                                        N
                                                        Q
                                                        M
                    VAQVDALSEHVFTNVPKCRDKTLNRSKM
   58 C
        SQKNVACKNGQTNCYQSYSTMSITDCRETGS
                                                S 90
   COO⁻ – VSADFHVPVYPNGECAVIIHKNAQTTKYACNPYK
      124

Fig. 2.4 Primary structure (sequence of amino acids) of protein on the example of ribonuclease A (RNase A), the enzyme digesting RNA. Each letter corresponds to one amino acid (residue).

4. Quaternary structure: non-covalent interactions that bind multiple polypeptides into a single, larger protein.

2.3.1 Hydrogen Bonds

Hydrogen bond is a type of relatively weak chemical bond consisting mainly on electrostatic attraction, but without the ionization of molecules (Fig. 2.5). A bond is formed by three atoms: one hydrogen and two negatively charged atoms, usually nitrogen (N) or oxygen (O). The hydrogen atom is connected by a covalent bond with one of the negatively charged atoms, called the **donor**. The second atom is called the **acceptor**. The electric charge is distributed between the three atoms that form a bond: electronegative atoms have partially negative charge, and hydrogen atom has partially positive charge. The strength of hydrogen bond depends on the donor and acceptor and their surrounding. Hydrogen bonds are an important factor stabilizing the structure of macromolecules such as proteins and nucleic acids.

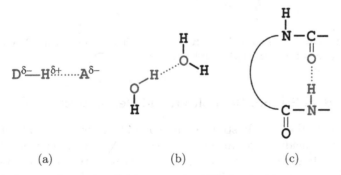

| (a) | (b) | (c) |

Fig. 2.5 Hydrogen bonds (denoted by dashed red line): (a) general formula; D: donor, A: acceptor; (b) the hydrogen bond between two water molecules; (c) the hydrogen bond in a polypeptide, e.g. α-helix or between two β-chains

2.3.2   Secondary Structure of Proteins

2.3.2.1   Alpha Helix

Alpha-helix is formed by amino acid chain in which every 3.6 amino acids make one turn. The distance between two turns is 0.54 nm. Shape of α-helix resembles a cylinder formed by tightly twisted chain (Fig. 2.6). The walls of the cylinder are created by polypeptide backbone when side chains (substituents) stick out of the cylinder. The conformation is stabilized by a periodic hydrogen bonds between amine group and an oxygen atom of the carbonyl group within the backbone of one polypeptide. α-helix can be right- or left-handed, although the right-handed form is the common one. If one side of the helix contains mainly hydrophilic amino acids, and the other — hydrophobic, α-helix has amphipathic properties (containing both polar and nonpolar portions in the structure). Among types of local structure in proteins, the α-helix is the most regular and the most prevalent, as well as the most predictable from sequence.

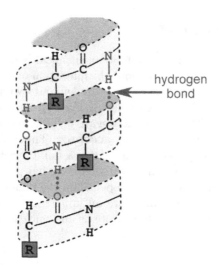

hydrogen
bond

Fig. 2.6 Scheme of regular α-helix. R — side chains of amino acids Altered from: http://www.answers.com/topic/alpha-helix

2.3.2.2   Beta Strand, Beta Sheet, and Beta Barrel

Beta-strand is almost fully stretched polypeptide fragment, usually a length of 5-10 amino acids. R groups of two neighboring amino acids are directed in the opposite directions. β-sheet is composed from two or more β-strands (Fig. 2.7). Its structure is stabilized by hydrogen bonds between amine groups and oxygen atoms from carbonyl groups of two neighboring chains (such way of spatial arrangement of amino acids resembles a piece of paper folded into a harmonica).

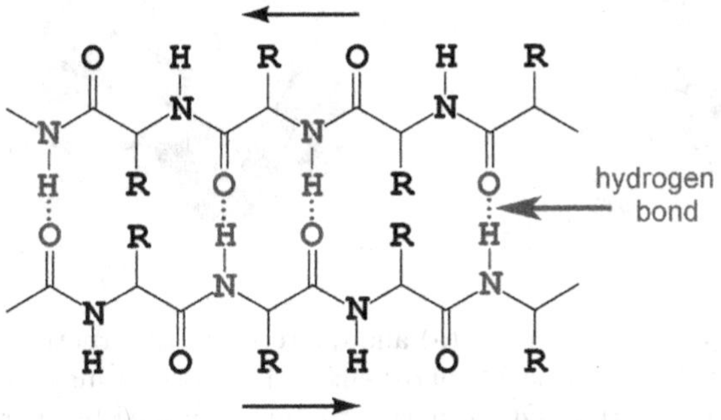

Fig. 2.7 Scheme of β-sheet with two anti-parallel β-strands connected by hydrogen bonds. Altered from: `http://en.wikipedia.org/wiki/File:BetaPleatedSheetProtein.png`

Beta-sheet can have parallel, anti-parallel or mixed arrangement. These structures (in contrast to the α-helix constructed from one continuous sequence) could consist from separate chain fragments, sometimes of different primary structure. Big β-sheets can form enclosed structures, in which the first chain connects with the last chain by hydrogen bonds. In this way the structure of β-barrel is created.

Secondary structure of protein can be predicted on the basis of primary structure. For this purpose programs available on websites can be used. Some examples are:

`http://bioinf.cs.ucl.ac.uk/psipred/`
`http://compbio.soe.ucsc.edu/SAM_T08/T08-query.html`
`http://distill.ucd.ie/porter/`
`http://www.predictprotein.org/`
`http://sable.cchmc.org/`

2.3.3 Tertiary and Quaternary Structure of Proteins

Tertiary structure describes the spatial organization of protein at a higher level than the secondary structure. Tertiary structure is the overall spatial distribution and interrelationships of turned chains and each amino acid residues in a single polypeptide chain (Fig. 2.8a). It is characterized by high compactness and it has no empty spaces inside of the molecule. Hydrophilic side chains and/or chains with a charge are typically on the surface of the molecule and they interact with water molecules. Large hydrophobic groups are generally hidden inside the protein molecule.

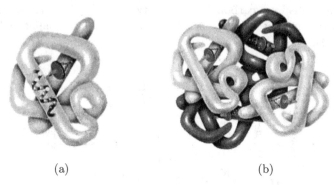

(a)   (b)

Fig. 2.8 An example of a tertiary (a) and quaternary (b) structure of the protein. A single peptide with conformation of α-helix and/or β-sheet may additionally fold, forming a tertiary structure. Quaternary structure is formed by aggregation of two or more peptides. Sourced from: `http://chsweb.lr.k12.nj.us/mstanley/ outlines/organicAP/aporgchem.html`

If the protein is composed from several polypeptides (so-called oligomer), a structure that is formed is called quaternary structure (Fig. 2.8b). The structure of the proteins is examined mainly by using X-ray crystallography and nuclear magnetic resonance (NMR).

Tertiary and quaternary structures of proteins are stabilized by different types of interactions between side chains of amino acids (hydrogen bonds and ionic bonds, hydrophobic interactions, covalent disulfide bonds).

2.4   Proteins Functions

Proteins are macromolecules with the greatest degree of diversity. It was estimated that in eukaryotic cells dozens-hundreds of thousands of different types of proteins are present. The structure of these proteins determines the functions they perform in the cell. The most important functions of proteins are:

- enzymatic catalysis — from the hydration of carbon dioxide to chromosome replication;
- immune reaction — e.g. immunoglobulins;
- regulatory and signaling (e.g. hormones and growth factors and their receptors that regulate biochemical processes and signal transduction in cells and between cells);
- control of membrane permeability — regulation of metabolite concentrations in cell;
- cell adhesion — e.g. catenin and cadherin;
- transport and storage of other molecules — e.g. hemoglobin, transferring, ovalbumin and casein;

- ordered movement (for example muscle contraction) — e.g. actin, myosin;
- building and structural function — e.g. keratin, elastin, collagen.

Beside classification based on function, proteins could be classified by location in the living cell. According to this, it is possible to classify all proteins into four main groups: membrane or transmembrane proteins, internal proteins, external or secretory proteins, and virus proteins. Another classification, based on posttranslational modifications, split all proteins into overlapped groups:

- native proteins (not changed after translation),
- glycoproteins (modified by covalent binding with linear or branched oligosaccharides),
- cleaved proteins (cleaved into two or more pieces),
- proteins with disulphide bonds (pair of cysteins are linked between each other by disulphide bond, called also disulphide bridge),
- protein complexes,
- chemically modified,
- prions.

Some main protein functions are described below.

2.4.1 Enzymes

Enzymes are proteins that catalyze (accelerate) biochemical reactions in the cell. It is the best-known role of proteins. Almost all chemical processes in living organisms (and viruses) require the participation of enzymes to provide adequate efficiency of the reaction. Enzymes are highly specific to substrates and, therefore, one enzyme catalyzes only a few reactions from many possible for given substrates. In this way enzymes determine metabolic and biochemical processes associated with the functioning of living organisms.

Enzymes, like other catalysts, accelerate the reaction by decreasing the energy barrier (activation energy) between the substrate (S) and product (P) (Fig. 2.9). The enzyme (E) needs a small portion of energy for the reaction because it forms transient connection with the substrate (S). This connection is called the enzyme-substrate complex (ES). The reaction proceeds as follows:

SUBSTRATE + ENZYME ↔ COMPLEX E-S → PRODUCT + ENZYME

Based on catalyzed reactions, enzymes were divided into following classes:

1. **Oxidoreductases** (e.g. dehydrogenases, oxidases, reductases, catalases) — catalyze the oxidation and reduction reactions.
2. **Transferases** (e.g. acetyltransferases, methylases, kinases) — catalyze the transfer of functional groups (respectively: acetyl, methyl, phosphate groups; Fig. 2.10).

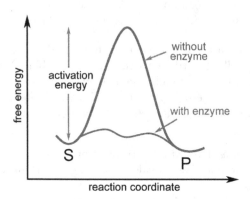

Fig. 2.9 Reduction of the activation energy in a reaction catalyzed by an enzyme

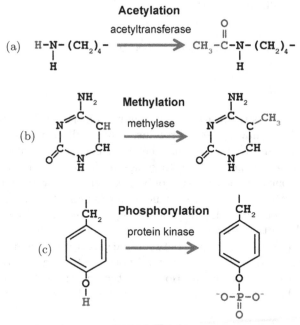

Fig. 2.10 Examples of the main types of chemical reactions proceeded by enzymes: (a) acetylation — addition of an acetyl group, -C(=O)CH3, catalyzed by acetyltransferase, e.g. to the side chain of amino acid (in this case — lysine); (b) methylation — attachment of a methyl group, -CH3, catalyzed by methylase, e.g. to cytosine in DNA chain; (c) phosphorylation — attachment of a phosphate group, -H2PO4, catalyzed by protein kinase, e.g. to the tyrosine side chain (serine and threonine may also be phosphorylated). Substituted groups are marked with red.

3. **Hydrolases** — catalyze hydrolysis reactions in which the molecule is split into two or more smaller parts with the participation of water. They include:

   - **Proteases** — digest protein molecules; e.g. caspases which play a crucial role in apoptosis;
   - **Nucleases** — digest nucleic acids: DNA is digested by DNase, and RNA by RNase. Exonucleases cut the single nucleotides from the end of the DNA or RNA molecule, endonucleases cut inside the molecule;
   - **Phosphatases** — catalyze dephosphorylation (removal of a phosphate group).

4. **Lyases** (e.g. decarboxylase and aldolase) — catalyze cutting C-C, C-O, C-S and C-N bonds in a manner other than hydrolysis or oxidation.
5. **Isomerases** (e.g. epimerases, racemases) — catalyze the atomic rearrangements within the molecule.
6. **Ligases** (e.g. synthetases, DNA ligase and RNA ligase) — catalyze the reaction of two molecules joining.

Enzymes can be inhibited by different types of molecules. Some of these molecules are competitive inhibitors, competing with substrates for reversible binding to the active center of the enzyme. The non-competitive inhibitors bind firmly to the enzyme molecules, permanently "removing" part of enzyme from the system. Enzyme inhibitors are often used as drugs. The best known enzymes which are targets of therapy are:

- HIV protease — essential for the replication of HIV (the inhibitor is used in treatment of AIDS);
- angiotensin converting enzyme (ACE) — causes contraction of blood vessels (an ACE inhibitor is used in treatment of high blood pressure);
- HMG-CoA reductase — an essential for the synthesis of cholesterol (its inhibitors, e.g. statins, reduce cholesterol levels);
- cGMP phosphodiesterase — catalyzes the conversion of cGMP (cyclic GMP) to GMP (inhibited by Viagra, which is used for the treatment of erectile dysfunction).

2.4.2 Membrane Proteins

Membrane protein is a protein associated with a cell or an organelle membrane. It was estimated that more than half of all cellular proteins interact with membranes. Many of them are an integral part of the phospholipid membrane (transmembrane proteins or permanently embedded in the membrane from one side). The peripheral membrane proteins are transiently associated with the phospholipid membrane or with integral membrane proteins.

Structural proteins existing in the membrane are attached to the microfilaments of the cytoskeleton. This provides stability of the cell. Proteins present on the cell surface allow cells to recognize and interact with each other. These proteins are involved for example in the immune response (ch. 2.4.2.1). Membrane enzymes produce a variety of substances essential for the cell to function. Membrane receptors allow the transmission of signals between the cell and the outside environment (ch. 2.4.2.2). Membrane transport protein play an important role in maintaining the proper concentration of ions. These proteins are classified into two groups: the carrier proteins and the channel proteins (ch. 2.4.2.3).

2.4.2.1   Proteins Involved in Immune Response

An immune system protects an organism against disease. In order to function properly, it must detect a wide variety of foreign antigens (i.e. substances that induce an immune response, what results in production of antibodies), and distinguish them from the organism's own healthy tissue. In this process, proteins belonging to the immunoglobulin family are involved. Molecules are categorized as members of this family based on shared structural features with immunoglobulins (also known as antibodies); they all possess a domain known as an immunoglobulin domain. The antibodies produced by B cells (lymphocytes), present in the peripheral blood, do not have a transmembrane domain and are soluble secretory proteins. However, many proteins transferring signals associated with the immune response have one or two domains penetrating the cell membrane (transmembrane) (Fig. 2.11). A typical immunoglobulin (for example antibody) is composed from four protein chains: two heavy (H) and two light (L). Variable region in both chains, directed outside of the cell, is denoted by the letter V (V_H and V_L domains). The more constant region is denoted by C_H and C_L. Variable region is an antigen binding site. The signal inside the cell is usually transmitted by a protein tyrosine kinase (an enzyme transferring a phosphate group on tyrosine), which is located on the cytoplasmic side of the cell membrane (ch. 8.4.1).

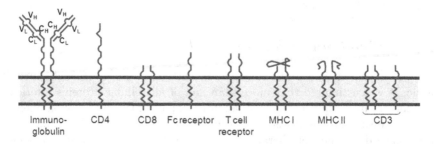

Fig. 2.11 Schematic representation of the structure of the immunoglobulin family proteins

2.4.2.2   Cell Surface Receptors

Many membrane proteins are cell surface receptors. They take part in communication between the cell and the outside environment. Extracellular signaling molecules attach to the receptor, triggering changes in the function of the cell. This process is called signal transduction (ch. 8). Membrane receptors are often classified based on their molecular structure or membrane topology. The polypeptide chains of the simplest are predicted to cross the lipid bilayer only once, while others cross as many as seven times, for example, the so-called G protein-coupled receptors (Fig. 2.12). They are a large group of proteins. They could be activated by agonists or blocked by antagonists. Photons, fragrances, pheromones, opiates, low molecular weight compounds such as amine, amino acids, ions, peptides and proteins, lipids and others could bind to the G protein-coupled receptor (ch. 8.3).

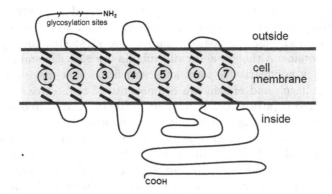

Fig. 2.12 Schematic arrangement of a membrane receptor which associates with G protein (β-adrenergic receptor)

Membrane receptors include also enzyme-linked receptors (usually single-pass transmembrane proteins; ch. 8.4) and ion channel linked receptors (multipass transmembrane proteins; ch. 8.2).

2.4.2.3   Ion Channels

Ion channels are created by a protein or proteins forming pores in the cell membrane. There are over 300 types of ion channels in a living cell. Their function is to generate and sustain a small potential gradient between the inner and outer surface of the cell membrane. Ion channels are present on all membranes of living organisms. Passage of ions through some of ion channels is dependent on the ion charge only. Pore diameter in such channels is just one or two atoms wide at its narrowest point. In some channels, passage through the pore is governed by a gate, which may be opened (activated) or closed (inactivated) by chemical or electrical signals, temperature, or mechanical force. Ion channels may be classified by the nature of their gating, the species of ions passing through those gates or the number of gates (see also ch. 8.2).

Proper activity of ion channels is particularly important in the nervous system, where they enable the propagation of nerve signals through synapses. Some predators produce toxins paralyzing the nervous system of their victims by affecting ion channels (venoms produced by e.g. spiders, scorpions and snakes). Ion channels are also a key component of the biological processes that require rapid changes in the cell, such as muscle contraction, transport of nutrients and ions, the activation of T cells, and insulin secretion from pancreatic beta cells.

2.4.3 Structural Proteins

Structural proteins are less "active" than those involved in catalyzing reactions, signaling cells, and transporting molecules. They confer stiffness and rigidity to otherwise-fluid biological components. Most structural proteins are fibrous proteins. Although actin and tubulin are globular and soluble as monomers, they polymerize to form long, stiff fibers that make up the cytoskeleton. Collagen and elastin are components of connective tissue (e.g. in the skin), and keratin builds hair, nails, feathers, hooves, and some animal shells. When the aging process begins the human body stops producing elastin. Without elastin replenishment, collagen begins to lose its elasticity and begins to weaken. It results in the loss of the skin's tone and resiliency. Thus, some companies advertise the use of collagen and elastin in their anti-aging creams and moisturizers; however, these products work only on the skin surface.

Structural proteins are also essential components of muscles (motor proteins such as myosin, kinesin, and dynein as well as actin), and are necessary to generate the force which allows muscles to contract and move. This class of proteins is also called contractile proteins.

References

1. Berg, J.M., Tymoczko, J.L., Stryer, L.: Protein Structure and Function. In: Biochemistry, 5th edn., W.H. Freeman, New York (2002)
2. King, M.W.: Basic Chemistry of Amino Acids. In: themedicalbiochemistrypage.org, LLC (1996-2012), http://themedicalbiochemistrypage.org/amino-acids.php
3. King, M.W.: Enzyme Kinetics. In: themedicalbiochemistrypage.org, LLC (1996-2012), http://themedicalbiochemistrypage.org/enzyme-kinetics.php

4. King, M.W.: Protein Structure and Analysis. In: themedicalbiochemistry-page.org, LLC (1996-2012), http://themedicalbiochemistrypage.org/protein-structure.php

5. Lodish, H., Berk, A., Zipursky, S.L., et al.: Protein Structure and Function. In: Molecular Cell Biology, 4th edn. W.H. Freeman, New York (2000)

6. Petsko, G.A., Ringe, D.: Protein Structure and Function. New Science Press Ltd., London (2004)

7. Vargas-Parada, L. (ed.): Proteins and Gene Expression. In: Miko, I. (ed.) Cell Biology. Nature Education (2011), http://www.nature.com/scitable/topic/proteins-and-gene-expression-14122688

Chapter 3
Nucleic Acids

Nucleic acids are polymers formed from nucleotides. Deoxyribonucleotides build DNA (deoxyribonucleic acid), and ribonucleotides build RNA (ribonucleic acid). With the exception of a few viruses, DNA forms the genetic material in all living organisms. Whereas, the major role of RNA is participation in protein synthesis.

3.1   Nucleotides

Nucleotide consists of three parts: a monosaccharide containing five carbon atoms (pentose), a nucleobase, and a phosphate group (Fig. 3.1). In nucleotide present in DNA and RNA there is one phosphate group linked with pentose, whereas free nucleotides can include two (e.g. ADP — adenosine diphosphate) or three (e.g. ATP — adenosine triphosphate) phosphate groups. Nucleotide,

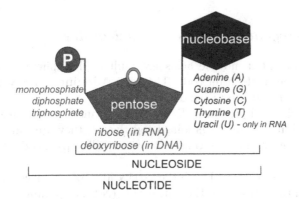

Fig. 3.1 The basic structure of the nucleotide

in which all phosphate groups are removed (i.e. only the nucleobase and pentose) is called a **nucleoside**.

3.1.1 Pentoses

Pentoses are monosaccharides with five carbon atoms. Saccharides (carbohydrates) are built of carbon chains with side hydroxyl groups (-OH) and a carbonyl group ($>C=O$) of aldehyde type (aldose) or ketone type (ketose). Ribose is an aldopentose, which is present in nucleic acids in the form of a ring (Fig. 3.2). Ribose is a component of RNA. The DNA contains deoxyribose, which means that in DNA ribose is lacking an oxygen atom at carbon 2 (2'-deoxyribose). Because of that RNA is called "ribonucleic acid" and DNA — "deoxyribonucleic acid".

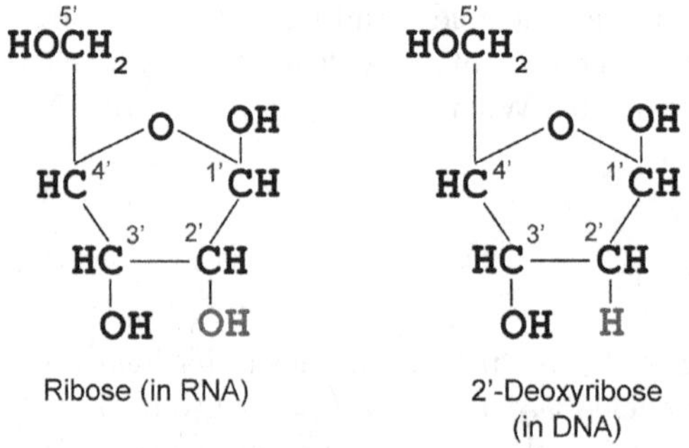

Fig. 3.2 The chemical structure of pentose ring forms: five carbon atoms are labeled from the C1' to C5'. In RNA **ribose** is present, in DNA — **deoxyribose**, which lacks an oxygen atom at C2'.

3.1.2 Nitrogenous Bases of Nucleotides

Nucleotides present in nucleic acids are built of five different nucleobases denoted with the single letter code. They are: **Adenine** (A), **Cytosine** (C), **Guanine** (G), **Thymine** (T) and **Uracil** (U).

A, C, G and T are present in DNA; A, C, G and U are present in RNA. A and G are built of two rings joined together — they are called **purines**. C, T and U contain only one ring — they are called **pyrimidines** (Fig. 3.3 and 3.4).

Nucleobases are connected with the saccharides by a N-glycosidic bond formed between the nitrogen in the purine ring (N-9), or pyrimidine ring (N-1) and the hydroxyl group at carbon 1 (C-1') in pentose. Table 3.1 gives the names of existing bases and their derivatives.

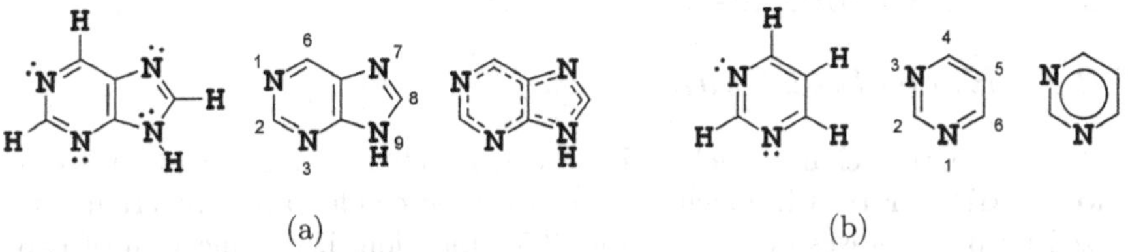

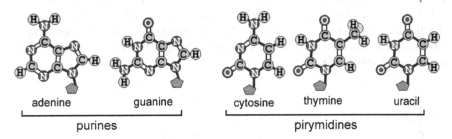

(a)   (b)

Fig. 3.3 General formulas of purine (a) and pyrimidine (b)

adenine   guanine   cytosine   thymine   uracil

purines   pirymidines

Fig. 3.4 The chemical structure of nucleobases found in DNA and RNA. Green pentagon shows the location of pentose.

Table 3.1 The names of the bases, nucleosides, and nucleotides

BASE	NUCLEOSIDE	RIBONUCLEOTIDES (5'-MONOPHOSPHATE)
Adenine (A)	adenosine	adenosine monophosphate (AMP), adenylic acid or adenylate
Guanine (G)	guanosine	guanosine monophosphate (GMP), guanylic acid or guanylate
Cytosine (C)	cytidine	cytidine monophosphate (CMP), cytidylic acid or cytidylate
Uracil (U)	uridine	uridine monophosphate (UMP), uridylic acid or uridylate
BASE	DEOXYNUCLEOSIDE	DEOXYRIBONUCLEOTIDES (5'- MONOPHOSPHATE)
Adenine (A)	deoxyadenosine	deoxyadenosine monophosphate (dAMP) or deoxyadenylate
Guanine (G)	deoxyguanosine	deoxyguanosine monophosphate (dGMP) or deoxyguanylate
Cytosine (C)	deoxycytidine	deoxycytidine monophosphate (dCMP) or deoxycytidylate
Thymine (T)	thymidine	(deoxy)thymidine monophosphate (dTMP) or (deoxy)thymidylate

3.2   DNA Properties

3.2.1   Nucleotide Chain

In the nucleotide chain, nucleotides are connected with each other by a **phosphodiester bond**, which is formed by the condensation reaction catalyzed by polymerases or ligases (Fig. 3.5). Ligation, i.e. connection of two chains of nucleic acids is catalyzed by ligases, while the synthesis of a nucleic acid chain (by attaching nucleotides) is catalyzed by RNA polymerase or DNA polymerase. The phosphate group, forming phosphodiester bonds in the chain of nucleic acids, binds with hydroxyl groups at carbons 3' and 5' of adjacent pentose rings (there are also possible other types of phosphodiester bonds in nucleotides, such as an intramolecular bond in so-called cyclic monophosphate; Fig. 8.2).

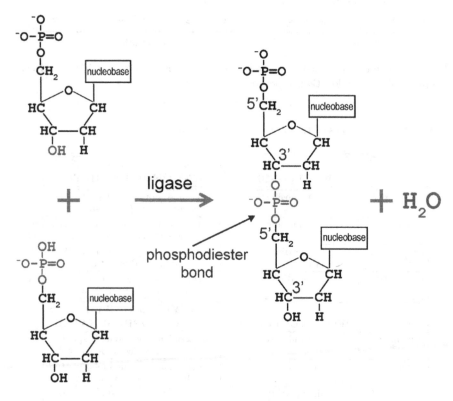

Fig. 3.5 Formation of a phosphodiester bond by condensation reaction

Like the peptide chain, a nucleic acid chain also has the orientation resulting from asymmetric character of bonds. 5' end has a free phosphate group

at carbon C5' of pentose, 3' end has a free hydroxyl group at carbon C3'
of pentose (Fig. 3.6). Synthesis of the nucleic acid chain always proceeds in
the direction from 5' to 3'. Because of that, unless it is stated otherwise, the
sequence of nucleotides in the nucleic acid chain is written from the 5' to
3' end (from left to right). The sequence of nucleobases along the backbone
encodes information. This information is read using the genetic code (ch. 3.5),
which specifies the sequence of the amino acids within proteins. The code is
read by copying stretches of DNA into the related nucleic acid RNA in a
process called transcription (ch. 5).

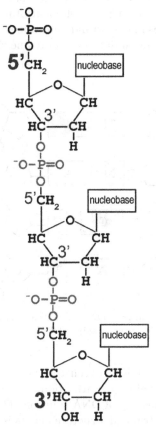

Fig. 3.6 The nucleic acid chain. 5' end has a free phosphate group (at C5'), the
3' end — a free hydroxyl group (at C3').

The DNA molecule is usually built from two chains, while most molecules
of RNA - from a single chain. Length of chain is defined by the number
of bases (or nucleotides) in the chain. In case of double-stranded nucleic
acid nucleobases from neighboring chains create pairs. In this case, the chain
length is given as the number of **base pairs (bp)**. Short chains of nucleic

acids (less than 50 nucleotides) are defined as oligonucleotides. Longer chains are polynucleotides.

3.2.2 DNA Structure

The two chains of the DNA molecule are held together by hydrogen bonds between the "paired" nitrogen bases. Adenine forms two hydrogen bonds with thymine and cytosine forms three bonds with guanine (Fig. 3.7). Although other pairs (such as G:T or C:T) also may form hydrogen bonds, they are weaker than those in pairs present in DNA (C:G and A:T). Thanks to the specific base pairing, the DNA strands are complementary with each other. Thus, the sequence of nucleotides in one strand determines the sequence of complementary strand. That is why in databases only one strand sequences (in the orientation 5' to 3') are given.

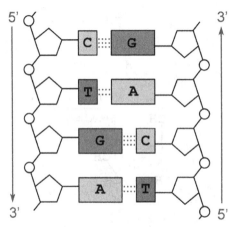

Fig. 3.7 Schematic base-pairing in DNA. Adenine (A) always forms a pair with thymine (T) and guanine (G) with cytosine (C)

Breaking of hydrogen bonds between the complementary bases leads to separation of two DNA chains (denaturation, melting). Such effect can be achieved for example by heating of DNA solution. The temperature at which loss of half of the double-stranded structure occurs is called the melting temperature (T_m). When the temperature is lowered, the DNA strands are again reconnected complementarily. This DNA property is used in hybridization techniques.

Two polynucleotide chains in a DNA molecule run in opposite directions to each other (i.e. are arranged anti-parallel). They are spinned around the common axis forming double-stranded helix. Purine and pyrimidine bases are located inside the helix (forming hydrophobic core), while the phosphate groups

and deoxyriboses — outside (forming hydrophilic, polar, external backbone).
Double helix, first described in 1953 by James Watson and Francis Crick, is
called **B form** of DNA. In that form helix is right-handed, a full turn of the
helix takes 3.4 nm and the distance between two neighboring pairs of bases
is 0.34 nm. Intertwined strands create two grooves of different width, called
the major and minor groove (Fig. 3.8). Such a structure facilitates binding
of specific proteins to DNA.

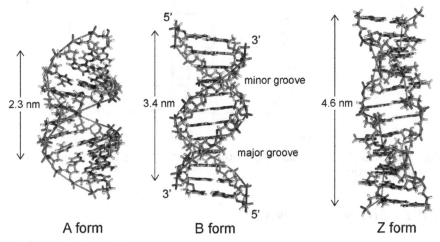

A form   B form   Z form

Fig. 3.8 The structures adopted by the DNA double helix. Sourced from:
http://en.wikipedia.org/wiki/DNA

In high salt concentration solutions or after the addition of alcohol, the
structure of DNA is converted to **A form**. It is also right-handed, but full
turn takes place every 2.3 nm, so one turnover occurs every 11 base pairs.
The A form in the cell may be produced in hybrid pairings of DNA and RNA
strands, as well as in enzyme-DNA complexes. Segments of DNA where the
bases have been chemically modified by methylation may undergo a larger
change in conformation and adopt the **Z form**, which phosphate backbone
resembles a zigzag. It is a left-handed form of DNA: one turnover takes 4.6 nm
and occurs every 12 base pairs. Such structure of DNA can be recognized by
specific Z-DNA binding proteins and may be involved in the regulation of
transcription (Fig. 3.8).

3.3   Structure and Function of RNA

RNA is synthesized in the cell in the process of transcription on the basis of
information contained in the sequence of DNA. Most RNA molecules retain
a single-stranded form and have a much shorter chain of nucleotides than the
DNA molecule. However, RNA can also form secondary structures such as

hairpin structure or stem-loop structure (Fig. 3.9). Main classes of RNA are **mRNA** (messenger RNA), **tRNA** (transfer RNA) and **rRNA** (ribosomal RNA). The main function of these classes of RNA is participation in protein synthesis. Moreover, in the cell there are present so called non-coding RNA (ncRNA), which is not involved in the translation process in a classical way. What is interesting, a number of RNA molecules shows, like proteins, catalytic activities. Such enzymatic RNA molecules are called ribozymes.

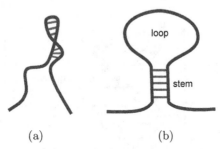

(a)   (b)

Fig. 3.9 Examples of secondary structures of RNA: (a) hairpin structure; (b) stem-loop structure

3.3.1 Messenger RNA

Messenger RNA (mRNA) carries coding information (copied from DNA sequence), which is used later for protein synthesis (translation) on ribosomes (ch. 6.1). The coding sequence of the mRNA determines the amino acid sequence in the protein that is produced. Three consecutive nucleotides (**codon**) in the mRNA encode an amino acid or form a stop signal for protein synthesis (Fig. 3.10).

```
                         codons
5′ – ATG TCA GAG GTG AAA TGC TAT GGT – 3′ DNA coding strand
3′ – TAC AGT CTC CAC TTT AGC ATA CCA – 5′ DNA template strand
                    |  transcription
5′ – AUG UCA GAG GUG AAA UGC UAU GGU – 3′ mRNA
                    |  translation
  N – Met Ser Glu Val Lys Cys Tyr Gly – C peptide chain
```

Fig. 3.10 The order of events in the synthesis of proteins: on the DNA template mRNA is synthesized in the process of transcription, on the mRNA template proteins are further produced in the translation process. mRNA sequence is complementary to DNA's template strand and identical to the coding strand (except that in RNA T is replaced by U). Commonly, in the case of DNA sequence only the sequence of the coding strand is given.

3.3.2 Transfer RNA

Transfer RNA (tRNA) consists of 74 - 95 nucleotides. It takes a complicated secondary structure similar to the cloverleaf, where approximately half of the nucleotides are paired (Fig. 3.11). Some of nucleotides in the tRNA molecule are methylated or modified in the other way, for example inosine (I), pseudouridine (ψ) or dihydrouridine (D). The main task of tRNA is translation of a sequence of nucleotides in mRNA into the sequence of amino acids in the peptide. This is possible because in the middle loop of tRNA the anticodon is located, which is complementary to the codon present on mRNA. 3' end of tRNA carries amino acid appropriate to ascribed codon (specific aminoacyl-tRNA is formed) (ch. 6.1.2).

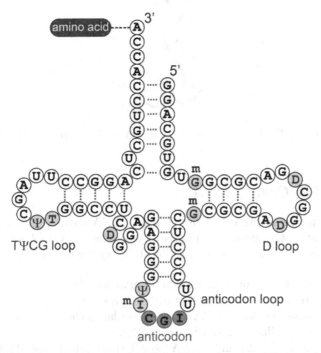

Fig. 3.11 Secondary structure of tRNA. Unusual or modified nucleotides are marked in blue (m - methylated). Anticodon is a triplet of nucleotides complementary to codon in the mRNA. When amino acid binds to the 3' end of tRNA, the formed molecule is called aminoacyl-tRNA.

3.3.3 Ribosomes and rRNA

In prokaryotic cells ribosomal RNA (rRNA) of three types (23S, 5S, and 16S) are present. In mammalian cells there are four types of rRNA: 28S, 5.8S, 5S, and 18S (unit "S" stands for Svedberg and it represents measure of sedimentation rate). rRNA is produced in the nucleus and transported to the cytoplasm, where it creates complexes with specific proteins, forming ribosome. Ribosome in 2/3 is composed of RNA and 1/3 of protein. It has two subunits: large and small. During protein synthesis the ribosome binds mRNA and tRNA, as shown in Fig. 3.12. Only tRNA which contains the anticodon matching to mRNA codon (located centrally between units of the ribosome) may connect to the complex.

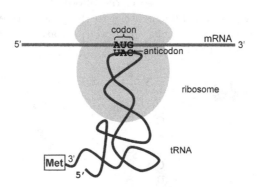

Fig. 3.12 The mRNA – ribosome – aminoacyl-tRNA complex formed during synthesis of protein

3.3.4 Non Coding RNA

In addition to the mentioned above classes of RNA participating directly in the process of protein synthesis, a range of other classes of RNA are produced in cells. They are called small non-coding RNA molecules. These include:

– snRNA (small nuclear RNA); low molecular weight nuclear RNA with a length up to 300 nt — involved in mRNA processing;
– snoRNA (small nucleolar RNA); low-molecular nucleolar RNA — responsible for the modification of ribosomal RNA;
– miRNA (micro RNA) and siRNA (small/short interfering RNA) — involved in gene expression regulation.

In recent years, the special interest is above siRNA molecules, due to the fact that they are routinely used in laboratories around the world to silence gene activity at the mRNA level. Both **miRNA** and **siRNA** have a length of 18 to 25 nucleotides, and their effect in the cell is similar (Fig. 3.13). There is consensus in the literature for the definition of miRNA and siRNA. Generally name miRNA is reserved for the molecules produced endogenously (i.e. inside the cell), siRNA — for delivered to the cell from outside. miRNA is formed from larger precursor (pri-miRNA), transcribed in the nucleus by polymerase

RNA II (which is also involved in transcription of mRNA). Such transcript is trimmed and transported to the cytoplasm, where it arrives in the form of approximately 70-nucleotide pre-miRNA forming stem-loop structure. The loop is removed leaving miRNA duplex. Base pairing in miRNA duplex is not perfect, so miRNA does not have 100% sequence homology to the target mRNA. siRNA is formed either from dsRNA (double standed RNA) or from shRNA (small/short harpin RNA) coming from viruses, transposons or delivered to cells experimentally. Like in the case of miRNA, initially siRNA duplex is formed. It has 100% homology to sequence of the target mRNA. miRNA or siRNA duplexes are then separated to single strands. One strand is removed, the other inhibits the translation of complementary mRNA (miRNA) or initiates its degradation (siRNA). As a result expression of the gene is diminished (encoded protein is not created). Inhibition of expression associated with the presence of miRNA is likely to occur also through the induction of epigenetic gene silencing (by methylation of histones, which leads to chromatin condensation).

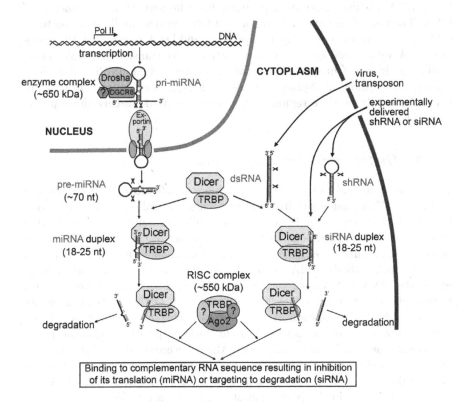

Fig. 3.13 miRNA and siRNA metabolism in the cell. Altered from:
http://www.nature.com/cr/journal/v15/n11/fig_tab/7290371f1.html

The phenomenon of silencing or exclusion of gene expression by double-stranded RNA with a sequence similar to the target DNA sequence is called **RNA interference (RNAi)**. It has an important role in defending cells against parasitic genes — viruses and transposons. On the other hand, miRNA are involved in negative regulation of gene expression during development. miRNA are involved in regulation of about 30% of human genes.

For the discovery of RNA interference in 1998, american researchers Andrew Z. Fire and Craig C. Mello received in 2006 Nobel Prize in Physiology or Medicine.

3.4   The Organization of Genetic Material in the Nucleus - Chromatin

Genomic DNA in the eukaryotic nucleus is hierarchically packaged into chromatin. Chromatin is a fibrous substance, which at the time of cell division becomes visible in the form of chromosomes (ch. 1.3.1). It is composed of DNA, RNA, histones, and other proteins. Each human cell has 1.8 meters of DNA. The role of chromatin is to package DNA into a smaller volume to fit in the nucleus and to control gene expression and DNA replication. Processes like DNA repair, replication, recombination, and transcription take place in the chromatin. Chromatin structure is changed dynamically: in regions where transcription takes place it is less condensed, loose (**euchromatin**), in transcriptionally inactive regions it is more condensed (**heterochromatin**) (Fig. 3.16).

3.4.1   Histones and Nucleosomes

The basic structural element of chromatin is **nucleosome**, which consists of 146 bp DNA and eight **histone** molecules (so-called histone octamer; Fig. 3.14). Histones are highly conserved alkaline proteins closely associated with DNA. They are responsible for the structure of chromatin and are involved in the regulation of gene expression. Nucleosome consists of 2 copies of each of histone proteins H2A, H2B, H3, and H4. They form the core, around which the DNA is wrapped forming two loops. The fifth type of histone, histone H1, which is more loosely connected with DNA, binds nucleosomes and is involved in formation of more complex chromatin structures. Interactions between histones and DNA occur between positively charged lysine side chains (present in large amounts in the N-terminus of histone) and negatively charged phosphate groups in DNA. If lysine is acetylated, its charge is neutralized. This leads to weakening of histones binding with DNA and plays a crucial role in the regulation of transcription (ch. 5).

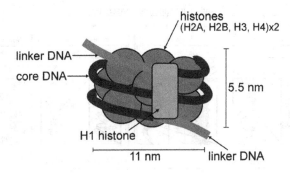

Fig. 3.14 Nucleosome structure

3.4.2 Chromatin Structure

In condensed chromatin each nucleosome binds with the next one by linker histone H1. This leads to the creation of 30 nm chromatin fiber (solenoid) (Fig. 3.15). In transcriptionally active chromatin, histone H1 is released from DNA and nucleosomes can change position, which means they move along the DNA strand. Nucleosome which never changes the position is called positioned nucleosome.

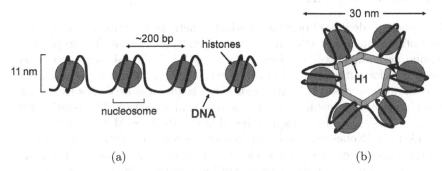

Fig. 3.15 Stages of chromatin organization: (a) nucleosomes fiber ("beads-on-a-string"); (b) single turn of solenoid (30 nm fiber)

Solenoid forms further more complex structures ensuring multilevel organization of chromatin in the nucleus. Important level of organization of chromatin are so called domains or loops, which are formed by interactions between nucleosome fibers and protein structures in nucleus (nuclear skeleton). The highest level of compaction of chromatin are chromosomes condensing during cell division (Fig. 3.16). The compacted DNA molecule is 40,000 times shorter than an unpacked.

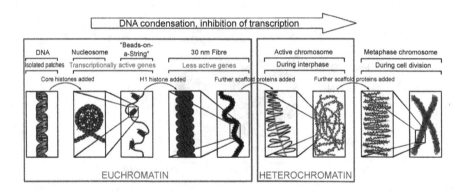

Fig. 3.16 Model of chromatin compaction and accompanying changes in gene activity. Altered from. http://en.wikipedia.org/wiki/ File:Chromatin_Structures.png

Recently is has been shown that 30 nm fibers are not detected in chromosomes *in situ* (i.e. examined exactly in the cell). It was hypothesised that crowding by cytoplasmic macromolecules is a sufficient factor that enables the packing of metaphase chromosomes.

3.5 Genetic Code

The genetic code is a set of rules, by which genetic information encoded in the nucleotide sequence of DNA or RNA may undergo translation into proteins (amino acid sequences) by living cells (Fig. 3.17). Translation takes place based on information stored in mRNA, which is built from nucleotides, while proteins are composed of amino acids. That is why it is necessary to link the exact nucleotide sequence with the amino acid sequence. The relationship between these sequences was deciphered in the 1960s by Marshall Nirenberg and his team (Nobel Prize in Physiology or Medicine in 1968). They correctly assumed that three successive nucleotides (codon) code for one amino acid.

The most important characteristic of the genetic code is its unambiguity, which means that a particular codon may encode only one type of amino acid. Genetic code is not strictly defined — an amino acid may be encoded by several different codons (we say that the genetic code is degenerated). If amino acid is encoded by several codons, the first two positions are usually the same, and differences exist at the third position. Moreover, amino acids with similar physical characteristics have similar codons. That is why point mutations (ch. 4.4) in DNA at the third position of codon usually does not change the sequence of encoded protein. Mutations at the first or second position more frequently change the amino acid, but more than likely it will have similar properties.

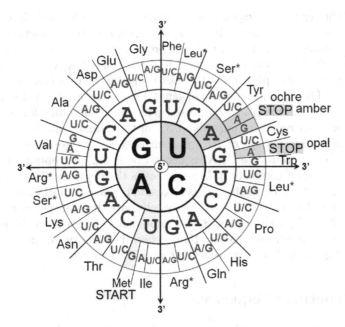

Fig. 3.17 The standard genetic code (codons should be read starting from the central ring). A peptide synthesis always begins with methionine (Met) encoded by the codon **AUG**, which encodes also methionine located inside the peptide chain. Stop codons (UAA, UAG, UGA) terminate peptide synthesis. In the diagram, nucleotides typical for RNA are given (in DNA U should be replaced by T). The DNA sequence from a start codon (ATG) to a stop codon (TAA, TAG or TGA) is called an **open reading frame** (ORF), which encodes a protein. Amino acids marked by ∗ can be encoded by several codons varying at the first nucleotide (for example arginine, Arg, can be encoded by AGA, or AGG as well as by CGA, CGG, CGU or CGC; different codons for the same amino acid can even appear side by side in the same molecule of mRNA).

Genetic code is universal for almost all organisms. There are however several exceptions to the rules of the standard genetic code. For example in the mitochondrial DNA of many organisms or in the nuclear DNA of several lower organisms there are different start or stop codons.

3.6 Genes and Genome

The definition of a gene has changed over time, starting with the definition introduced by Gregor Mendel in the nineteenth century. One of the most popular, although not the actual definition, was that a gene is a fragment of DNA encoding a protein chain. Currently it is assumed that the **gene** is an entire sequence of the DNA necessary to built a protein or RNA. The sequence of the gene can be divided into the **regulatory region** and the **transcribed**

region. The regulatory region can be distant or can be located directly next to the region of transcription. In eukaryotic organisms the transcribed region usually contains exons and introns (in prokaryotic organisms genes do not have introns). Exons encode functional proteins or RNA, while introns are removed after transcription (ch. 5). In DNA, in addition to genes, pseudogenes are present. They are non-functional genes. Often they are formed as a result of duplications and mutations of normal genes. Regions coding for functional proteins or RNA represent only about 3% of the human genome.

All genetic information of an organism is called its **genome**. For most organisms, it is a complete DNA sequence (Table 3.2). Exceptions are RNA viruses, in which a complete RNA sequence is their genome. With the term of genome, the term of **proteome** is associated, which comprises all proteins present in a cell at a given time. All body cells have the same genome, but can differ significantly with proteome. The proteome is larger than the genome, as there are many more proteins than genes. This is due to an alternative splicing of genes and post-translational modifications (ch. 5.8).

3.7   Repetitive Sequences

DNA sequences containing most genetic information in the form of structural genes are described as unique DNA. They are found in the genome in two copies - alleles located in corresponding parts of homologous chromosomes. However, many DNA sequences repeat several times in the total cellular DNA. For example, in the telomeres located at the ends of chromosomes, the sequence "TTAGGG" repeats thousands of times throughout about 15,000 bp. Such sequences are called **repetitive sequences**. The function of repetitive

Table 3.2 Variation in genome size and gene content

Organizm type	Organism	Genome size (Mb*)	Number of genes
Viruses	Hepatitis D	0.0017	2
	Hepatitis B	0.0032	4
	HIV-1	0.0092	9
	Bacteriophage λ	0.0485	90
Bacterium	*Escherichia coli*	4.6	4,437
Yeast	*Saccharomyces cerevisiae*	12.1	6,300
Nematode	*Caenorhabditi elegans*	100	19,000
Insect	*Drosophila melanogaster* (fruit fly)	130	14,000
Mammals	*Mus musculus* (house mouse)	3,000	20,000-30,000
	Homo sapiens (human)	3,200	20,000-30,000

* 1 Mb = 1 million base pairs (in the case of double-stranded DNA or RNA) or 1 million bases (in the case of single-stranded DNA or RNA).

DNA is largely unknown. The repetitive DNA may be clustered at one place or interspersed with unique sequences, each repeat unit ranging from a few hundred to few thousand base pairs.

Classification of repetitive sequences was originally based on kinetics of reassociation of denatured DNA. The total DNA is firstly randomly digested into fragments of average size of about 1,000 bp. Then the DNA fragments in aqueous solution are heated and complementary strands dissociate to single strands (due to breaking of hydrogen bonds). Decreasing of the temperature is connected with reassociation of DNA sequences (hybridization) according to the rule of complementarity of bases. Reassociation kinetics can be estimated for example by measuring of changes in absorbance of the solution at 260 nm. The absorption coefficient of the double-stranded DNA is about 40% lower than of single-stranded DNA. If DNA fragment contains a sequence that repeats several times in the genome, it has a greater chance to find a matching strand and reassociate faster than other fragments containing less repetitive sequences. Relying on the rate of reassociation, the DNA sequences were divided into three classes:

1. Highly repetitive: hybridizing very quickly (just a few seconds). In mammals they make up 10-15% of the DNA. Short sequences called microsatellites and tandem repeats are classified here.
2. Moderately repetitive: hybridizing in the intermediate speed. Represent 25-40% of the genome of mammals. Among these are interspersed repeats, also known as mobile elements or transposons.
3. Single or of small number of repetitions: represent 50-60% of mammalian DNA.

Interspersed repeats can be a short sequences of length 100-500 bp, called SINES (Short Interspersed Nuclear Elements) or long repetitions, with a length of several thousand base pairs, called LINES (Long Interspersed Nuclear Elements). An example of a sequence belonging to SINES are highly homologous Alu sequences with a length of 300 bp. In human genome they are present at level close to a million copies. They are dispersed throughout the genome and can serve as initiation sites of DNA replication. Another class of repetitive DNA often occurs in the centromeres (satellite DNA). In *Drosophila* satellite DNA is composed of seven-nucleotide sequence repeated more than 10,000 times. The genes encoding the 5.8S, 18S, and 28S rRNA appear in groups located in the nucleolus, and are repeated tandemly many times. The genes encoding histones are also found in tandemly repeated groups.

References

1. Berg, J.M., Tymoczko, J.L., Stryer, L.: DNA, RNA, and the Flow of Genetic Information. In: Biochemistry, 5th edn. W.H. Freeman, New York (2002)
2. Clancy, S.: Chemical structure of RNA. Nature Education 1(1) (2008), http://www.nature.com/scitable/topicpage/chemical-structure-of-rna-348

3. Clancy, S.: RNA functions. Nature Education 1(1) (2008),
   http://www.nature.com/scitable/topicpage/rna-functions-352
4. Hancock, R.: Structure of metaphase chromosomes: a role for effects of macromolecular crowding. PLoS One 7(4), e36045 (2012)
5. King, M.W.: Basic Chemistry of Nucleic Acids. In: themedicalbiochemistry-
   page.org, LLC (1996-2012), http://themedicalbiochemistrypage.org/
   nucleic-acids.php
6. Maeshima, K., Hihara, S., Eltsov, M.: Chromatin structure: does the 30-nm
   fibre exist in vivo? Curr. Opin. Cell Biol. 22, 291–297 (2010)
7. Lodish, H., Berk, A., Zipursky, S.L., et al.: Nucleic Acids, the Genetic Code,
   and the Synthesis of Macromolecules. In: Molecular Cell Biology, 4th edn. W.H.
   Freeman, New York (2000)
8. Phillips, T.: Small non-coding RNA and gene expression. Nature Education 1(1)
   (2008), http://www.nature.com/scitable/topicpage/
   small-non-coding-rna-and-gene-expression-1078
9. Pray, L.: Eukaryotic genome complexity. Nature Education 1(1) (2008),
   http://www.nature.com/scitable/topicpage/
   eukaryotic-genome-complexity-437
10. Yeung, M.L., Bennasser, Y., Le, S.Y., et al.: siRNA, miRNA and HIV: promises
    and challenges. Cell Research 15, 935–946 (2005)

Chapter 4
DNA Replication, Mutations, and Repair

4.1 DNA Replication Mechanism

Replication (i.e. copying) of DNA takes place before cell division. During replication one double-stranded DNA molecule produces two identical copies. Each strand of the original double-stranded DNA molecule serves as a template for the production of the complementary strand, a process referred to as semiconservative replication. Replication is triggered by a transcription factor, in yeast called MCB binding factor, in mammals — E2F, that regulates the expression of enzymes necessary for replication: DNA polymerases, DNA primases and cyclins. Replication starts in a specific place (called **replication origin**) (Fig. 4.1). In genomic DNA of *E. coli* there is only one place of replication origin (called *oriC*), whereas eukaryotic DNA has many such sites on each chromosome. In a sequence of replication origin, A:T pairs dominate, which are easier to separate than a C:G pairs, because they are connected only by two hydrogen bonds. Unwinding of DNA at the origin and synthesis of new strands leads to the formation of a replication fork (Fig. 4.2).

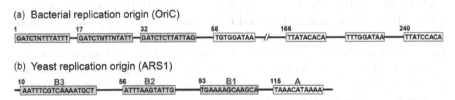

(a) Bacterial replication origin (OriC)

(b) Yeast replication origin (ARS1)

Fig. 4.1 Sequences of bacterial and yeast replication origin: (a) in bacteria it is a 245 bp sequence containing three 13 bp tandemly repeated segments of almost identical sequence (green), and four 9 bp segments with high sequence similarity (yellow); (b) yeast replication origin sequence is called ARS (autonomously replicating sequence). The A region is essential for the initiation of replication, while B1, B2 and B3 enhance efficiency of replication.

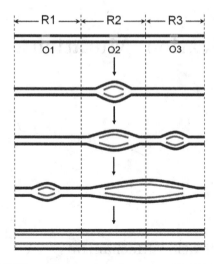

Fig. 4.2 Scheme of DNA replication in eukaryotes. O1, O2 and O3 — replication origins for replicons R1, R2 and R3. Replication requires unwinding of double helix. New strands are synthesized by DNA polymerase on template of old strands. Unwinded DNA forms replication fork, moving in one direction, thus "replication bubble" increases. Replication begins at some replication origins earlier than at others. As replication nears completion, "bubbles" of newly replicated DNA meet and fuse, finally forming two new molecules.

Replication occurs on both DNA strands simultaneously. *E. coli* needs 42 minutes to replicate the whole genomic DNA. The speed of the replication fork progression, formed by unwinded DNA, is about 1,000 bp per second. In eukaryotes, it moves much slower — only 50-100 bp per second. This is due to the association of DNA with histones. Replication of the entire human genome takes approximately 8 hours, *Drosophila* genome — only 3-4 minutes (thanks to the 6,000 replication forks operating simultaneously on a DNA molecule).

A DNA molecule is synthesized by DNA polymerases from deoxyribonucleoside triphosphates (dNTPs). The process is similar to the synthesis of RNA (ch. 5). Both polymerases, DNA and RNA, synthesize the nucleic acid chain only in the 5' to 3' direction. Because two chains of DNA are anti-parallel, only one strand (leading strand) can be synthesized by DNA polymerase continuously. The second strand (lagging strand) is synthesized in fragments (Fig. 4.3). Cellular proofreading and error-checking mechanisms ensure near perfect fidelity for DNA replication.

4.1.1 DNA Polymerases

DNA polymerases are enzymes that carry out all forms of DNA replication. DNA polymerase synthesizes a new strand of DNA by extending the 3' end of an existing nucleotide chain, adding new nucleotides complementary to the

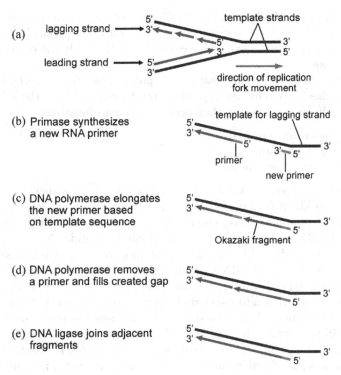

Fig. 4.3 Synthesis of lagging strand: (a) comparison of the leading strand and lagging strand; (b) primase synthesizes a new primer of a length of about 10 nt. The distance between two primers is about 1,000-2,000 nt in bacteria, and about 100-200 nt in eukaryotic cells; (c) DNA polymerase attaches nucleotides to the new primer in the direction 5' to 3' until it encounters the 5' end of the previous primer. Newly synthesized DNA fragment is called Okazaki fragment (named after their discoverer); (d) in *E. coli* DNA polymerase I shows exonuclease activity and from the 5' end removes encountered primer and fills created gap; (e) DNA ligase joins adjacent Okazaki fragments. The entire lagging strand is synthesized by repeating steps (b) to (e).

template strand. In *E. coli* there are three types of DNA polymerases: I, II and III. DNA polymerase I completes gaps between the fragments of DNA formed during the synthesis of the lagging strand. It is also involved in DNA repair. DNA polymerase II is activated in response to DNA damage (SOS response). The majority of DNA in prokaryotic cells is synthesized by DNA polymerase III. This enzyme is 900 kDa in size, and is composed of several subunits (they form so called holoenzyme). The polymerase core is made up of α, ε, and θ subunits. The role of other units is to prevent detachment of the enzyme from the template. Two β units (β_2) act as sliding DNA clamps, keeping the polymerase bound to the DNA. Together with the polymerase core they slide along the DNA molecule. It allows a continuous synthesis of approximately 5×10^5 nucleotides. In the absence of β subunits the polymerase core is

detached after synthesis of about 10-50 nucleotides. DNA polymerase III
can add about 1,000 nucleotides per second and it is extremely precise: ε
subunit, which is $3' \rightarrow 5'$ exonuclease, checks continuously the correctness of
the replication.

In mammals there are five main DNA polymerases ($\alpha, \beta, \gamma, \delta$ and ε) and
several smaller. γ polymerase replicates mitochondrial DNA, the others are
present in the nucleus, where they perform different functions:

– α: lagging strand synthesis;
– β and ε: DNA repair;
– δ: leading strand synthesis.

To start the synthesis of DNA a short RNA molecule (primer) is required.
The primer is about 5-12 nucleotides in length, and is complementary to one
strand of the template. It is synthesized by a specific enzyme — **primase**,
which is the RNA polymerase. The use of ribonucleotides for the synthesis
of primer means that this fragment is temporary — it is removed in the
final phase of replication. RNA polymerases, unlike DNA polymerases, can
synthesize new strand without a primer, because they do not check whether
the nucleotide inserted into the chain is correct (they are much less accurate).
Other proteins involved in replication are **helicases** and single-strand DNA
binding (SSB) proteins. Helicase unwinds double helix and SSB proteins bind
to single-stranded fragments, preventing from the duplex restoration.

The mechanism of replication in bacteria and eukaryotes is similar. However, eukaryotic DNA polymerases do not possess subunits similar for bacterial β subunit. When they attach to DNA, they use the PCNA protein
(**proliferating cell nuclear antigen**).

4.1.2 Polymerase Chain Reaction (PCR)

The ability of DNA polymerase to produce identical copies of DNA is utilized
in the laboratory to amplify fragments of DNA (the target sequence is used
many times as a template). Such strategy, named polymerase chain reaction
(PCR), is the most sensitive method to detect and amplify a target sequence.
Nowadays PCR is widely used for many applications, independently or in
combination with other techniques.

The polymerase can synthesize a new strand only on a single stranded template, thus the process takes place after DNA denaturation (see ch. 3.2.2).
In vitro, in a tube, it can be achieved by heating the DNA solution. Polymerases of most organisms cannot survive in high temperatures, however in
PCR, DNA polymerase from organisms living in hot springs is used, such as
Taq polymerase from *Thermus aquaticus*. To start the synthesis, the DNA
polymerase needs a primer. In PCR, two primers complementary to the target sequence are used (each complementary to a different strand). The sequence flanked by primers is amplified after consecutive PCR cycles (Fig. 4.4).

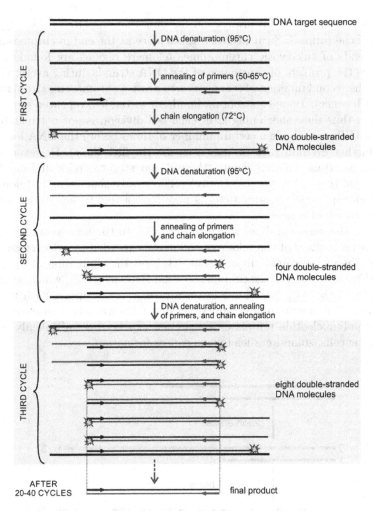

DNA target sequence

DNA denaturation (95°C)

annealing of primers (50-65°C)

chain elongation (72°C)

two double-stranded DNA molecules

FIRST CYCLE

DNA denaturation (95°C)

annealing of primers and chain elongation

four double-stranded DNA molecules

SECOND CYCLE

DNA denaturation, annealing of primers, and chain elongation

eight double-stranded DNA molecules

THIRD CYCLE

AFTER 20-40 CYCLES

final product

Fig. 4.4 Scheme of polymerase chain reaction

Products generated in one cycle are used as additional templates in the next cycle. Up to 40 cycles (consisting of three steps: denaturation, primers annealing and chain elongation) could be processed in PCR machine. After 30 cycles more than 1×10^9 molecules are already produced from one DNA molecule.

4.2 Telomerases and Cellular Senescence

Synthesis of lagging strand requires synthesis of multiple primers, that are finally removed. In bacteria, which have a circular DNA molecule, there is no problem with the replication of the whole molecule. In case of linear DNA molecule, lagging strand is always shorter than the template (at least by the length of the primer). Such situation takes place at the end of chromosomes. At the ends of eukaryotic chromosomes telomere regions are found, which minimize the problem of shortening of the DNA strands during every cell division (the second function of telomeres is to protect chromosomes from fusing with each other). Telomeres contain hundreds of tandem repeated sequences (ch. 3.7), thus their shortening after each cell division is not detrimental to the cell. Cells can divide a certain number of times before the DNA loss prevents further division (this is known as the Hayflick limit). However, after reaching a certain telemere length, the cell may stop to divide and die. This is a normal process in somatic (body) cells. Steady shortening of telomeres with each replication in somatic cells may have a role in senescence and in the prevention of cancer.

Within the germ cell line, which passes DNA to the next generation, the repetitive sequences of the telomere region are extended by telomerase. It is a large ribonucleoprotein enzyme that catalyzes the elongation of 3' end of DNA (Fig. 4.5). Telomerase contains internal RNA molecule, which serves as a template (Fig. 4.6). So telomerase is a reverse transcriptase with its own template. That is quite unique: any other of the known polymerases does not contain polynucleotide template. Telomerase can become mistakenly active in somatic cells, sometimes leading to cancer formation.

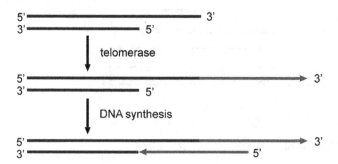

Fig. 4.5 Scheme of telomere elongation by telomerase. Telomerase first extends the longer strand of DNA. Then, using the same mechanism as synthesizing the lagging strand, the shorter strand is extended.

4.3   Topoisomerases

During DNA replication, DNA ahead of the replication bubble becomes positively supercoiled, while DNA behind the replication fork becomes entangled forming precatenanes (Fig. 4.7). Enzymes called topoisomerases play an essential role in resolving these topological problems.

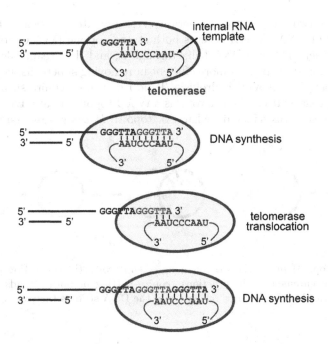

Fig. 4.6 The mechanism of telomere extension by telomerase. In human chromosome telomere has about 10-15 thousand base pairs, which consist of many tandemly repeated sequences GGGTTA. Telomerase contains RNA complementary to repetitive sequences in the telomere, which is used as a template for the synthesis of DNA. By repeating a cycle of DNA synthesis and telomerase translocation, telomere is extended by many repeats.

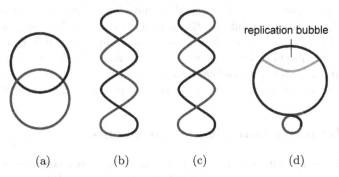

Fig. 4.7 The structure of catenane and supercoils of DNA: (a) circular catenane; (b) positive supercoil DNA (left-handed); (c) negative supercoil DNA (right-handed); (d) positive supercoil bacterial DNA during replication. Supercoil DNA molecule is more dense than the relaxed DNA of the same length.

There are two types of topoisomerases: type I creates a transient break in one strand of DNA, type II — cuts both strands of one DNA double helix, passes another unbroken DNA helix through it, and then reanneals the cut strand (Fig. 4.8). Type II is more efficient in removing supercoils, however it requires energy from ATP hydrolysis. Type I does not require such energy. Topo I in prokaryotes and eukaryotes is a type I topoisomerase. Examples of type II topoisomerases include eukaryotic topo II, *E. coli* gyrase and topo IV.

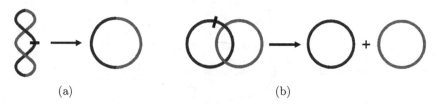

(a)   (b)

Fig. 4.8 Topo II functions: (a) removing of supercoils; (b) separation of linked molecules of catenanes. These actions require cutting of both strands of DNA (marked with black line). After relocation of the DNA strands they are again connected.

Eukaryotic topo I and topo II can relax both positively and negatively supercoiled DNA. In bacteria topo I catalyzes only the relaxation of negatively supercoiled DNA. Bacterial gyrase has two functions: relaxes positively supercoiled DNA during replication and forms a negative twists in the DNA (so it can be packed into a cell). During replication, these negative turns are removed by topo I.

Without topoisomerases DNA cannot replicate correctly. That is why inhibitors of topoisomerases are used as anticancer drugs that inhibit cell division. Such drugs inhibit the division of all cells in the body, including normal cells. Thus, for example hair loss is a common side effect of therapy (since hair are growing due to a permanent division of cells in hair follicles). DNA-damaging effect of topoisomerases inhibitors, outside of its potentially curative properties, may lead to secondary neoplasms in the patient.

4.4   Mutations and Their Consequences

A mutation is a permanent change in the DNA sequence concerning several nucleobases or changes on the scale of the chromosome. Effects of the mutation could be observed only when it occurs in gene exons or regulatory elements (ch. 5.3). Changes in the non-coding regions of DNA generally do not affect function. If a mutation occurs in a germ-line cell (i.e., egg or sperm cells), it can be passed to an organism's offspring. In contrast to germ-line mutations, somatic mutations occur in cells found elsewhere in an organism's

body. Such mutations are passed to daughter cells, but they are not passed
to offspring.

In dipliod organisms, which have two copies of each gene (alleles), mutations are recessive or dominant. **Recessive mutations** inactivate the affected gene and lead to a loss of function. Thus, to give rise to a mutant
phenotype, both alleles must carry the mutation (otherwise the correct allele
will provide a proper function). In contrast, the phenotypic consequences of
a dominant mutation are observed in a heterozygous individual carrying one
mutant and one normal allele. **Dominant mutations** often lead to a gain of
function. Mutations in one allele may also lead to a structural change in the
protein that interferes with the function of the wild-type protein encoded by
the other allele. These are referred to as dominant negative mutations.

Mutations can be advantageous and lead to an evolutionary advantage,
and can also be deleterious, causing disease, structural abnormalities, devel-
opmental delays, or other effects. In this chapter mutations occurring on a
small scale, such as substitutions, deletions, insertions, and exon skipping
resulting from mutations in splicing site are presented.

4.4.1 Substitution

Mutation called substitution occurs when one or more nucleotides are replaced by the same number of other nucleotides. In most cases it is only one
nucleotide (**a point mutation**). One purine, i.e. A or G, could be replaced
with another purine or one pyrimidine, i.e. C or T, with another pyrimidine
(such substitution is called a **transition**). Replacement of a purine with a
pyrimidine or vice versa is called a **transversion**. Transition mutations are
about an order of magnitude more common than transversions. Because of
the substitution consequences, they can be divided into silent mutations (not
leading to changes in the encoded protein), mutations leading to amino acid
change in the encoded protein (missense), and mutations inserting premature
stop codon (nonsense) (Fig. 4.9).

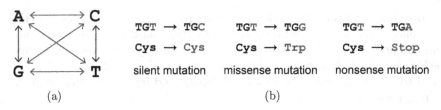

(a)   (b)

Fig. 4.9 Categorizing point mutations: (a) illustration of transition (blue arrows)
and transversion (red arrows), (b) functional categorization: examples of silent mutation, mutation leading to amino acid change in the encoded protein, and mutation
introducing a premature stop codon

4.4.2 Deletion and Insertion

Deletion is a mutation in which one or more nucleotides are removed from the DNA sequence. When three nucleotides are deleted (equivalent of the codon) one amino acid is lost in the protein chain, but the rest is not changed (Fig. 4.10a). The deletion which is not a multiple of three nucleotides (Fig. 4.10b, c, d) leads to shift of the reading frame and change of all codons starting from place of deletion (ch. 6.1.4).

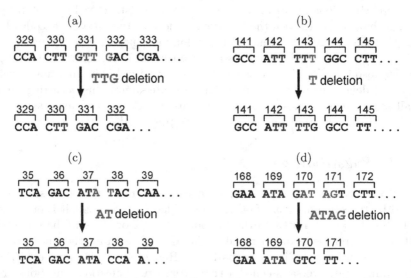

Fig. 4.10 Examples of deletions leading to the development of the disease: (a) deletion in the *F9* gene that could be connected with hemophilia B; (b) and (c) deletions in the *CFTR* gene that can cause genetic disorders (the blockage of the ions and water movement into and out of the cell); (d) deletion in the *APC* gene that could occur early in cancers, such as colon cancer. Sourced from: http://www.web-books.com/MoBio/Free/Ch7E.htm

Insertion consists in insertion of one or more extra nucleotides into the sequence. Similarly to deletion it can cause frameshift and formation of abnormal protein or, if inserted nucleotides are a multiple of three, new amino acids are added. Insertions which does not change reading frame also can lead to serious diseases (Table 4.1). Deletions and insertions often occur in repetitive sequences.

4.4.3 Exon Skipping

Cutting of intron requires the presence of "GU . . . . AG" signal (Fig. 4.11; ch. 5.8.4). If the splicing acceptor site AG is mutated (e.g. A to C),

Table 4.1 Examples of diseases caused by insertions

Disease	Affected gene	Repetitive sequence	Number of repeats normal	Number of repeats mutated	Direct consequence of mutation
Fragile X syndrome (common cause of autism and intellectual disability)	fragile X mental retardation 1 (*FMR1*) promoter	CGG	6-44	> 200	methylation of the promoter leading to the silencing of the *FMR1* gene
Huntington's disease (neu-rodegenerative disorder)	huntingtin (*HTT*)	CAG	9-35	37-100	expansion of the polyglutamine tract
Spinal and bulbar muscular atrophy (SBMA) (neuro-degenerative disorder)	androgen receptor (*AR*)	CAG	7-24	40-55	expansion of the polyglutamine tract
Spinocerebellar ataxia (SCA1) (neurodegenera-tive disorder)	ataxin 1 (*ATXN1*)	CAG	19-36	43-81	expansion of the polyglutamine tract
Dentatorubral-pallidoluysian atrophy (DRPLA) (spinoce-rebellar degeneration)	atrophin 1 (*ATN1*)	CAG	7-23	> 49	expansion of the polyglutamine tract
Duchenne muscular dystrophy	dystrophin (*DMD*)	CTG	5-35	50-4,000	premature translation termination

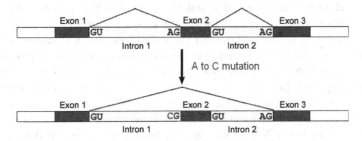

Fig. 4.11 Illustration of exon skipping

ribonucleoprotein complex cutting intron will search for the next acceptor place, which will result in excision of exon together with introns.

4.4.4 The Relationship between Mutations and Polymorphisms

DNA sequence variations are sometimes described as mutations and someti-mes as polymorphisms. A mutation could be defined as any change in a DNA sequence away from normal. In contrast, a **polymorphism** is a DNA sequ-ence variation that is common in the population. The arbitrary cut-off point between a mutation and a polymorphism is 1%. That is, to be classed as a po-lymorphism, the least common allele must have a frequency of 1% or more in the population. If the frequency is lower, the allele is regarded as a mutation.

The most common type of variation in the human genome is the single nucleotide polymorphism (SNP or "snip"), where a single base differs between individuals. SNPs occur about once every 1,000 — 2,000 base pairs in the genome. Polymorphic sequence variants usually do not directly cause diseases. Many are completely neutral in effect, some may influence characteristics such as height and hair colour rather than characteristics of medical importance. However, polymorphic sequence variation does contribute to disease susceptibility and can also influence drug responses.

4.5   Mutation Mechanisms

Mutations can be caused by external (UV light, chemical agents, viruses, etc.) or endogenous (toxic products of cellular metabolism) factors, or they may be caused by errors in the cellular machinery (replication errors, accidental deamination, etc.).

Deamination (Fig. 4.12) of cytosine to uracil in DNA is potentially mutagenic because uracil forms a pair with adenine. After replication, one of the daughter strands will contain the modified pair A:U instead of the correct pair G:C. However, since uracil is not the part of DNA, this mutation can easily be detected and repaired by base excision (ch. 4.6.1). Althought the chemical structure of uracil is simpler than the structure of thymine, in DNA thymine is used, not uracil. Otherwise, deamination of cytosine to uracil could be distinguished from properly inserted uracil only by the incorrect pairing with the complementary base. It would have to be corrected in the process of mismatch repair (ch. 4.6.3), which is much less effective than base excision. Thus, thymine is present in DNA because it increases the reliability of the genetic code. Unlike DNA, RNA can not be repaired, so uracil is used to RNA synthesis as "cheaper" element (its synthesis requires less energy than synthesis of thymine).

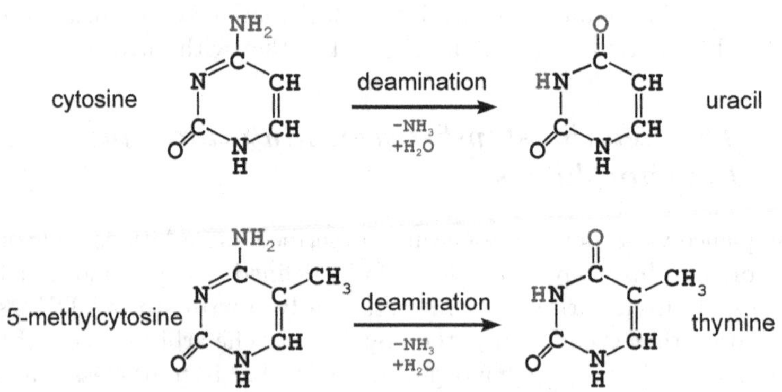

Fig. 4.12 Examples of deamination, i.e. the removal of the amine group. Accidental deamination can change the cytosine to uracil or the methylated cytosine to thymine.

Spontaneous deamination of 5-methylcytosine results in thymine (Fig. 4.12). This is the most common single nucleotide mutation in DNA. The repair mechanisms do not recognize thymine as erroneous, thus this change cannot be corrected. It coud be a reason for the rarity of CpG sites (ch. 4.5.4) in the eukaryotic genome.

4.5.1 Mutations Caused by UV Light

When DNA is exposed to UV light, two neighboring pyrimidines (cytosines or thymines) may form a dimer (Fig. 4.13). Fortunately, most of these genetic lesions are corrected seconds after they are created, before they can do permanent damage. However, unrepaired dimers are mutagenic (Fig. 4.14).

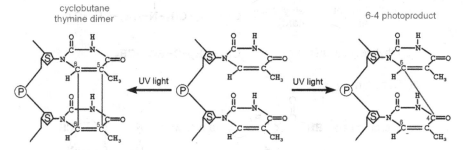

Fig. 4.13 Thymidine dimers induced by UV light (similar dimers may be formed from cytosine)

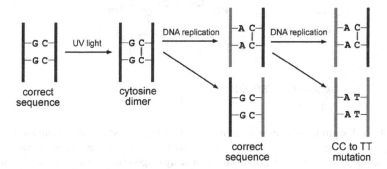

Fig. 4.14 Possible mechanism of mutation induced by UV light. If the cytosine dimer is formed by UV light, during replication in the new strand (light blue) adenine could be incorporated (instead of guanine). In the next round of replication opposite to the adenine, thymine will be built, which will cause the mutation consisting in a change of CC to TT. While cytosine dimers can be repaired, mutations arising after replication is no longer detectable by the DNA repair system.

4.5.2   Mutations Caused by Chemical Agents

Chemical agents that induce mutations in DNA are called **mutagens** and are said to be mutagenic. Most of them are also carcinogens. In figures 4.15 and 4.16 structures of several potential mutagens and the mechanism of mutation induced by some of them are shown.

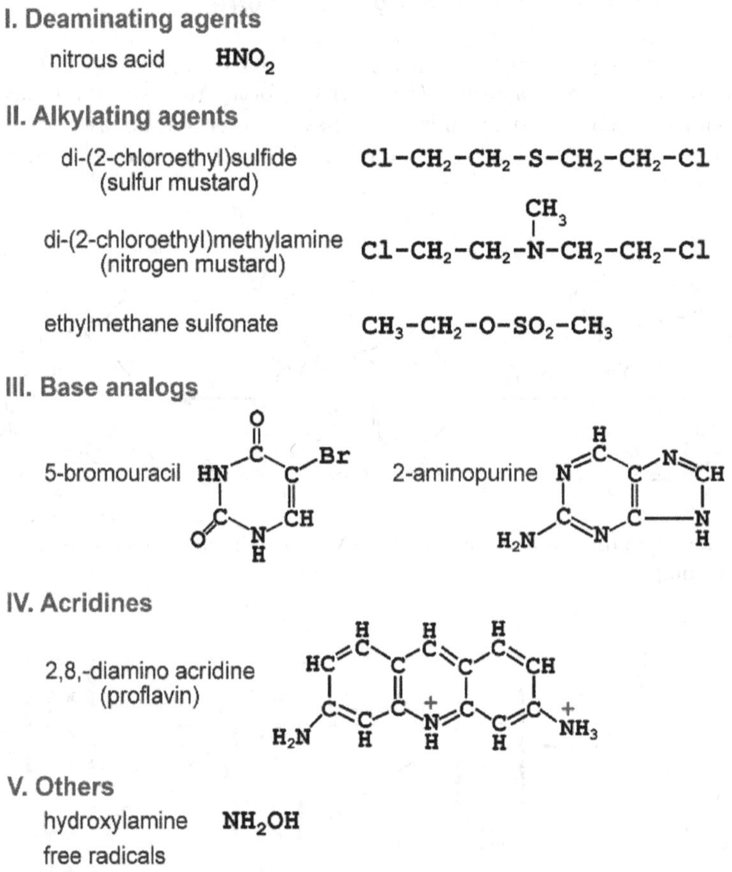

I. Deaminating agents

nitrous acid HNO_2

II. Alkylating agents

di-(2-chloroethyl)sulfide
(sulfur mustard) $Cl-CH_2-CH_2-S-CH_2-CH_2-Cl$

di-(2-chloroethyl)methylamine
(nitrogen mustard)

$$\overset{\displaystyle CH_3}{\underset{\displaystyle |}{Cl-CH_2-CH_2-N-CH_2-CH_2-Cl}}$$

ethylmethane sulfonate $CH_3-CH_2-O-SO_2-CH_3$

III. Base analogs

5-bromouracil

2-aminopurine

IV. Acridines

2,8,-diamino acridine
(proflavin)

V. Others

hydroxylamine NH_2OH
free radicals

Fig. 4.15 Examples of potential mutagens: I. Nitrous acid is a deaminating factor, which converts cytosine to uracil, adenine to hypoxanthine, guanine to xanthine; these bases have a different potential to form hydrogen bonds than the initial ones, which leads to changes in the pairing of bases; II. Alkylating agents which add alkyl group ($-C_nH_{2n+1}$) to other molecules; nucleobase alkylation in DNA may alter normal base pairing and thus cause mutations; some alkylating agents can also cross-link DNA molecules, causing breaks in chromosome; III. A base analog is a chemical that can substitute for a normal nucleobase in nucleic acids (5-bromouracil — for thymine, 2-aminopurine — for adenine); IV. Acridines (e.g. proflavine) are positively charged molecules; they can be inserted between two strands of DNA, which leads to disorders in DNA structure and replication; V. Hydroxylamine and free radicals alter the structure of bases, leading to changes in the base pairing.

4.5.3 Mutations Caused by Replication Errors

Errors are natural part of the DNA replication. Although cells employ an arsenal of editing mechanisms to correct mistakes, they do happen. They are the main source of mutations. The frequency of mutations that have not been corrected at the time of replication is one mutation per one cell division.

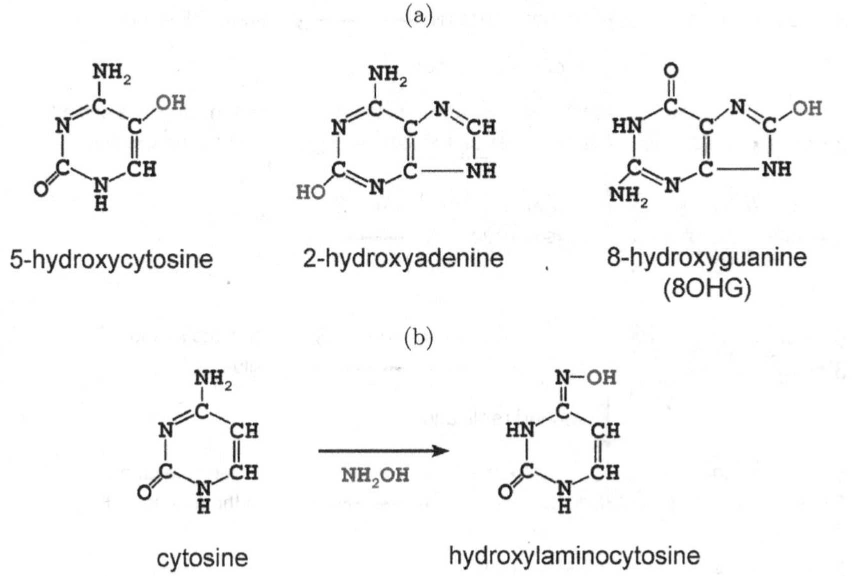

(a)

| 5-hydroxycytosine | 2-hydroxyadenine | 8-hydroxyguanine (8OHG) |

(b)

cytosine → hydroxylaminocytosine

Fig. 4.16 (a) Structures of nitrogenous bases formed under the influence of free radicals; (b) change of base induced by hydroxylamine (NH_2OH)

Most DNA replication errors are caused by mispairings of a different nature. For example in repetitive sequences, where a new strand may be incorrectly paired with the template strand, the replication slippage is common (Fig. 4.17). Such shifts formed during replication lead to the creation of polymorphisms, e.g. in the microsatellite sequences containing thousands of repetitive sequences. If the mutation occurs in a protein-coding sequence, it can lead to production of incorrect protein and cause a disease. Huntington's disease is a well known example (Table 4.1).

4.5.4 DNA Methylation and CpG (CG) Islands

The CG island is a genomic region, in which the frequency of the CG sequence is higher than in other regions. It is also called the CpG island, where "p" means that C and G are linked with each other by a phosphodiester bond. In vertebrates, CpG islands are often found in the promoters of genes

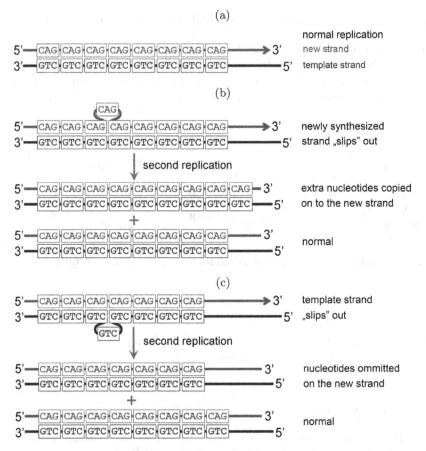

Fig. 4.17 Mutations caused by replication slippage within the repetitive sequence. In this figure wrong pairing involves only one repetition but there could be more of them. (a) Normal replication; (b) backward slippage, resulting in the insertion mutation; (c) forward slippage, resulting in the deletion mutation.

required for the maintenance of basic cellular function (so-called **housekeeping genes**), or genes that are frequently expressed. In the promoters of active genes, CpG sequences are not methylated, whereas in inactive genes - they are usually methylated, what is the reason of gene expression inhibition. The cytosines in the CpG tend to be methylated. This methylation helps to distinguish the newly synthesized DNA strand from the parental one, which is important in the final stages of DNA proofreading after duplication. However, methylated cytosine can be converted into thymine by the accidental deamination (Fig. 4.12). Such mutation can only be corrected by mismatch repair, which is very inefficient. Thus, over evolutionary time the methylated CG sequences tend to turn into the TG sequences, which explains the lack of CG sequences in inactive genes.

Each cell type has its own DNA methylation pattern, so that a unique set of proteins can be produced in the cell and enable it to perform functions specific for the cell type. During cell division, the DNA methylation pattern passes to the daughter cell. This happens due to DNA methyltransferase, which methylates only those CG sequences, which are paired with methylated CG (Fig. 4.18).

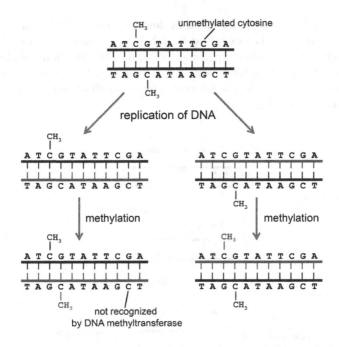

Fig. 4.18 Inheritance of the DNA methylation pattern. DNA methyltransferase methylates only those CpG sequences, which are paired with methylated CpG in parental strand. Thanks to this, after replication of DNA the original methylation pattern is preserved.

4.6 Mechanisms of DNA Repair

Most changes in DNA are quickly repaired. Those that are not repaired result in a mutation. Thus, mutation is, in fact, a consequence of the failure of DNA repair. There are three main repairing mechanisms of damaged or inappropriate bases: by base excision, by nucleotide excision, and by mismatch repair. Other harmful changes in DNA, like double-strand breaks, are repaired by non-homologous end joining (NHEJ) or homologous recombination. In NHEJ pathway, the broken ends are directly ligated without the need for a homologous template, in contrast to homologous recombination, which requires a homologous sequence to guide repair. Inappropriate NHEJ can lead to translocations and telomere fusion. Homologous recombination, beside DNA repairing, produces also a new combinations of DNA sequences during meiosis (ch. 7.4.1), the process by which eukaryotes make gametes, like sperm and egg in animals.

4.6.1 Base Excision Repair (BER)

Nucleobases in DNA can be modified by deamination or alkylation. Damaged or inappropriate base is recognized by DNA glycosylase, which removes it, forming **AP site** (apurinic/apyrimidinic site). The backbone of the DNA remains intact. AP site is then cleaved by an AP endonuclease. The resulting single-strand break can then be processed by either short-patch or long-patch base excision repair (Fig. 4.19).

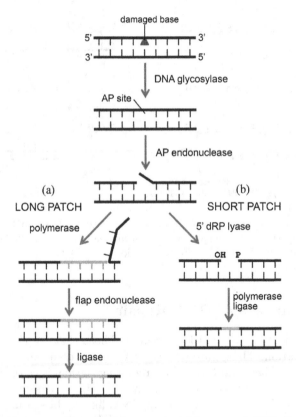

Fig. 4.19 Basics steps of base excision repair: (a) long patch BER (2-10 nucleotide patch); (b) short patch BER (one nucleotide patch)

4.6.2 Nucleotide Excision Repair (NER)

While the base excision repair machinery can recognize and correct only damaged bases, the nucleotide excision repair (NER) enzymes recognize bulky distortions in the shape of the DNA double helix. NER is a particularly important for removing of the UV-induced DNA damage (e.g. thymidine dimer). It is recognized differently depending on whether the DNA is transcriptionally active (transcription-coupled repair) or not (global excision repair) (Fig. 4.20). After the initial recognition step, the damage is repaired in a similar manner. A short single-stranded DNA segment (24-32 nucleotides), that includes the lesion, is removed. A single-strand gap created in the DNA is subsequently filled in by DNA polymerase, which uses the undamaged strand as a template.

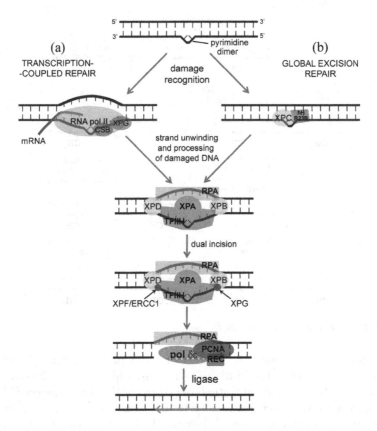

Fig. 4.20 DNA repair by nucleotide excision. NER can be divided into two subpathways that differ only in their recognition of helix-distorting DNA damage: (a) transcription coupled NER and (b) global genomic NER.

4.6.3   DNA Mismatch Repair (MMR)

The specificity of mismatch repair is primarily for base-base mismatches and insertion/deletion mispairs generated during DNA replication and recombination. The MMR machinery distinguishes the newly synthesised strand from the template (which is treated as a correct one). In eukaryotes, the exact mechanism for this is not clear. In gram-negative bacteria, a transient hemimethylation distinguishes the strands: immediately after synthesis only the template strand is methylated. In *E. coli* all adenines which in the template strand are in the context of the sequence 5'-GATC-3' are methylated by a special methylase (Dam methylase). The enzyme that removes mismatch cuts unmethylated strand, leaving the parental strand intact, so it could serve as a template during the repair of DNA (Fig. 4.21).

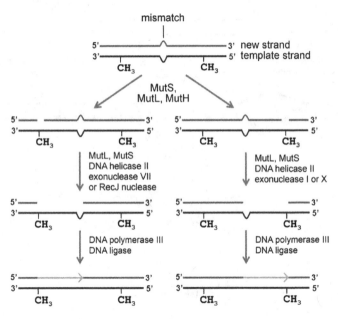

Fig. 4.21 The mismatch repair in *E. coli*

The process begins with the binding of MutS protein within the wrong paired base pairs. Then the MutL protein joins, which leads to binding and activation of MutH. It cuts unmethylated strand within the sequence GATC. The sequence from the place of incision to the wrong paired region is removed by exonuclease with assistance of helicase II. The exonuclease recruited is dependent on which side of the mismatch MutH incises the strand — 5' or 3'. If the nick is on the 5' end of the mismatch, either RecJ or ExoVII (both 5' to 3' exonucleases) is used. If however the nick is on the 3' end of the mismatch, ExoI (a 3' to 5' enzyme) is used. Formed gap is filled by

DNA polymerase III and DNA ligase with assistance of SSB (single stranded DNA-binding) proteins.

The distance between the GATC site and the area where there are mistakes in pairing rules could be up to 1,000 bp. Therefore, the mismatch repair system is expensive and not very efficient. This system works similarly in eukaryotic organisms. Proteins homologous to bacterial MutS and MutL were found in yeast, mammals, and other organisms. These are the MSH5 MSH1 (homologous to MutS) and MLH1, PMS1, and PMS2 (homologous to MutL). Mutations in the MSH2, PMS1, and PMS2 genes are associated with an increased risk of colon cancer.

References

1. Berg, J.M., Tymoczko, J.L., Stryer, L.: DNA Replication, Recombination, and Repair. In: Biochemistry, 5th edn. W.H. Freeman, New York (2002)
2. Clancy, S.: DNA damage & repair: mechanisms for maintaining DNA integrity. Nature Education 1(1) (2008), http://www.nature.com/scitable/topicpage/dna-damage-repair-mechanisms-for-maintaining-dna-344
3. King, M.W.: DNA Metabolism. In: themedicalbiochemistrypage.org, LLC (1996-2012), http://themedicalbiochemistrypage.org/dna.php
4. Lodish, H., Berk, A., Zipursky, S.L., et al.: DNA Replication, Repair, and Recombination. In: Molecular Cell Biology, 4th edn. W.H. Freeman, New York (2000)
5. Loewe, L.: Genetic mutation. Nature Education 1(1) (2008), http://www.nature.com/scitable/topicpage/genetic-mutation-1127
6. Pray, L.: DNA replication and causes of mutation. Nature Education 1(1) (2008), http://www.nature.com/scitable/topicpage/dna-replication-and-causes-of-mutation-409
7. Pray, L.: Major molecular events of DNA replication. Nature Education 1(1) (2008), http://www.nature.com/scitable/topicpage/major-molecular-events-of-dna-replication-413
8. Pray, L.: Semi-conservative DNA replication: Meselson and Stahl. Nature Education 1(1) (2008), http://www.nature.com/scitable/topicpage/semi-conservative-dna-replication-meselson-and-stahl-421
9. Zvereva, M.I., Shcherbakova, D.M., Dontsova, O.A.: Telomerase: structure, functions, and activity regulation. Biochemistry (Mosc) 75, 1563–1583 (2010)

Chapter 5
Transcription and Posttranscriptional Processes

5.1 Gene Expression Overview

In multicellular organisms all somatic cells have the same set of DNA, yet different cell types differ in shape and performed function. This is possible because different cells within the body express different portions of their DNA, what determines diversity. Genes do not have influence on cellular functions as long as they are not expressed.

Gene expression is the process by which information from a gene is used in the synthesis of a functional gene product (a protein or functional RNA). The process involves several steps (Fig. 5.1):

- transcription: on DNA template complementary RNA is synthesized (precursor RNA, **pre-RNA**);

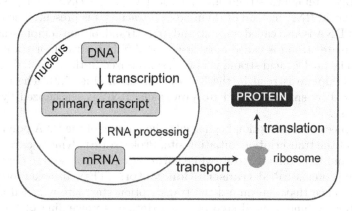

Fig. 5.1 Subsequent steps of the expression of genes encoding proteins in eukaryotic cells, where transcription and translation are separated in time and space. In prokaryotes translation can begin before the end of gene transcription.

– RNA processing: modifications of pre-mRNA leading to the formation of mature mRNA or a functional tRNA, rRNA, or ncRNA. In the case of genes encoding RNA their expression ends when a functional RNA molecules reach their destinations.

Expression of genes encoding proteins includes two additional steps:

– transport of mRNA to the cytoplasm;
– protein synthesis: in the cytoplasm mRNA binds to ribosomes where protein is synthesized on the basis of mRNA sequence.

According to the process described above, the flow of genetic information occurs in the following direction: DNA → RNA → protein. This rule has been called the **central dogma** of molecular biology because it was thought that the same rule would apply to all organisms. However, it is not true for RNA viruses, where the genetic information can be transferred back from the RNA to the DNA.

5.2 Gene Transcription and RNA Polymerases

Transcription is a process in which one DNA strand is used as a template for the synthesis of a complementary RNA. For example:

```
5' - GATGCAGTGAGCTCAGGATCTA - 3'    DNA coding strand (+)
3' - CTACGTCACTCGAGTCCTAGAT - 5'    DNA template strand (−)
5' - GAUGCAGUGAGCUCAGGAUCUA - 3'    RNA (+)
```

Thus, synthesized RNA strand has a nucleotide sequence complementary to the DNA template strand and identical with the DNA coding strand, except that in RNA instead of thymine (T) uracil (U) is present. The coding strand of DNA is also called sense strand, plus strand, or non-template strand. The template strand is called antisense strand, minus strand, or transcribed strand. The nucleic acid strand is always synthesized in the 5' to 3' direction, and the template is read in the 3' to 5' direction (Fig. 5.2). The reaction is catalyzed by enzymes called **polymerases**: RNA is synthesized by RNA polymerase.

In all species transcription begins with the binding of the RNA polymerase to DNA at the transcription initiation site. Prokaryotic polymerases recognize this site and bind directly to it, while eukaryotic polymerases are dependent on other proteins called **transcription factors** (TFs). In eukaryotes, the DNA sequence that determines the transcription start site is called a **core promoter**. A fundamental step of transcription is unwinding of the DNA fragment. The enzyme which unwinds the double helix is called a **helicase**. Prokaryotic polymerases possess also the helicase activity. In eukaryotes DNA is unwound by a specific transcription factor (TFIIH; ch. 5.5).

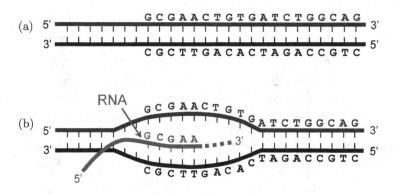

Fig. 5.2 Scheme illustrating the process of transcription: (a) DNA before the transcription; (b) during the transcription fragment of DNA is unwinded, so that one of the strand can be used as a template for the synthesis of a complementary RNA (so-called transcription bubble is formed)

Activated RNA polymerase moves along the template and adds matching RNA nucleotides that are paired with complementary DNA bases. Elongation involves a proofreading mechanism that can replace incorrectly incorporated bases. Transcript elongation leads to clearing of the promoter, and the transcription process can begin yet again. Thus, multiple RNA polymerases can work simultaneously on a single DNA template, and many RNA molecules (especially mRNA) can be rapidly produced from a single copy of a gene. RNA polymerase remains connected with DNA until it encounters on transcription termination signal. Prokaryotic and eukaryotic organisms have different signals for transcription termination. (Note: the signal "stop", which is present in the genetic code, is a signal to terminate protein synthesis, not transcription). Transcription in eukaryotic cells is much more complicated than in prokaryotes, partially because the eukaryotic DNA is associated with histone proteins that may block the access of polymerase to the promoter.

5.2.1 RNA Polymerases

Both RNA and DNA polymerases are enzymes that add nucleotides to an existing strand of nucleic acid, what leads to its elongation. **However, RNA polymerase can start a new strand synthesis without any primer.** RNA polymerase uses nucleoside triphosphates (NTPs). Synthesis starts from pppG or pppA (p — phosphate group). Nucleotide is attached to the RNA, while two phosphate groups in the form of pyrophosphate (PP_i) are released

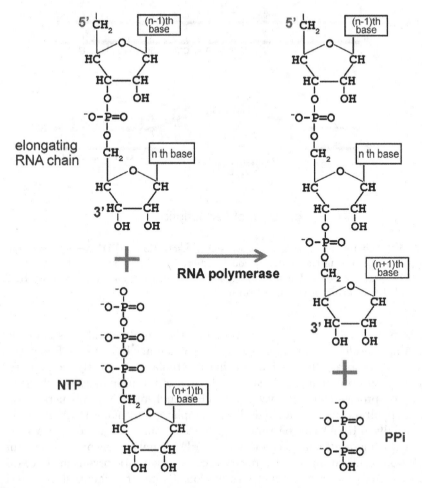

Fig. 5.3 Chemical reaction catalyzed by RNA polymerase

(Fig. 5.3). **Nucleic acid strand is always synthesized in the 5' to 3' direction**.

Prokaryotic (e.g. *E. coli*) RNA polymerase is a large (approximately 500 kDa) enzyme complex consisting of five subunits: four catalytic (two α, one β, one β') and a single regulatory (σ). Beta subunits (β, 151 kD, and β', 156 kD) are much larger than α subunit (37 kD). Sigma subunit (also called σ factor) appears in several forms, with a mass from 28 kD to 70 kD. It plays an important role in recognizing the transcription initiation site and also shows helicase activity. Catalytic subunits ($\alpha_2\beta\beta'$) form **polymerase core**. Together with σ subunit it is called **holoenzyme**. The term "holoenzyme" is generally applied to enzymes that contain multiple protein subunits, all needed for activity.

In eukaryotes there are three classes of RNA polymerases: I, II and III. Each of them consists of two large and 12 to 15 smaller subunits. Large subunits are homologs of prokaryotic β and β' subunits, and two smaller are similar to the α subunit of *E. coli*. Eukaryotic polymerases do not contain subunits corresponding to σ factor in *E. coli*. In eukaryotes other proteins are involved in the initiation of transcription.

Polymerase II is the most important among the eukaryotic RNA polymerases. It is involved in the transcription of protein-coding genes, snRNA and miRNA. The other two polymerases are responsible only for the transcription of genes encoding RNA. RNA polymerase I is located in the nucleolus and it synthesizes rRNA (except 5S rRNA). RNA polymerase III is located outside the nucleolus and it synthesizes 5S rRNA, tRNA, U6 snRNA and some small RNA.

5.3 DNA Regulatory Elements

In the simplest term, a gene is a region of transcription together with a regulatory region (Fig. 5.4). Transcription unit is a sequence of DNA undergoing transcription (i.e., a sequence which serves as a template for complementary primary transcript). In the regulatory region, **cis-regulatory elements** can be distinguished. These are specific DNA sequences, to which **transcription factors (TFs)** may bind. TFs are proteins, which after binding to DNA affect transcription (enchance or suppress it). DNA sequences that encode transcription factors are called **trans-regulatory elements**.

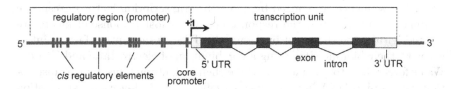

Fig. 5.4 The structure of a typical gene coding for protein in eukaryotes. Dark rectangles within the transcription unit - coding sequences; bright rectangles - sequences coding for untranslated regions (UTRs) present in the 5' and 3' ends of the mature transcript. Nucleotides in the sequence downstream from the transcription start site (indicated by an arrow pointing to the right) are numbered strating from +1, and upstream - from -1 (no 0 position).

With the regulatory region the terms "basic (core) promoter", "enhancer", and "silencer" are also linked. The core promoter is formed from DNA sequences determining transcription start site (abbreviated to TSS; also called Inr — initiator). The **enhancer** is a sequence, which after binding of the transcription factor (in this case — the activator) increases the activity of the promoter. The **silencer**, after binding of the transcription factor (repressor), inhibits transcription. Sometimes the same sequence can enhance or inhibit transcription, depending on the binding protein. Enhancers and silencers could be found upstream or downstream from the transcription unit, or even in introns.

In eukaryotes the most common element of the promoter of protein-coding genes is the **TATA box**, found 20-35 bp upstream from the transcription start site (at position -35 to -20). TATA box is nothing more than a sequence TATAAA (sometimes: TATATAT or TATATAA). The TATA box is the binding site for a transcription factor known as TATA-binding protein (TBP)(ch. 5.5). Upstream by 200 bases to the TSS, regulatory elements like **CAAT box** or **GC box** commonly occur (Table 5.1).

Table 5.1 Sequences commonly found in promoters of eukaryotic genes and transcription factors that bind to them

Name of the sequence	Position	Transcription factors	Consensus sequence
TATA box	-35 to -20	TATA-box binding protein (TBP)	TATAAA
CAAT box	-110 to -50	CAAT binding protein (CBF), nuclear factor 1 (NF1), or CAAT/enhancer binding protein (C/EBP)	GGCCAATCT
GC box	-200 to -70	Specificity protein/Krüppel-like factor (SP/KLF) family	GGGCGG

5.3.1 Response Elements

The DNA sequence recognized by a specific transcription factor in response to some stimuli is called the response element. Most response elements are located within one thousand bp above the TSS. Transcription factors interact with their binding sites using a combination of electrostatic and Van der Waals forces. Due to the nature of these chemical interactions, most transcription factors bind DNA in a sequence specific manner. However, transcription factors do not bind just one sequence but are capable of binding a subset of closely related sequences, each with a different strength of interaction. The most common sequence recognized by a given transcription factor is called consensus, or logo sequence.

Table 5.2 describes some of the eukaryotic response elements. CRE (cAMP response element) interacts with CREB (CRE-binding) protein, which is regulated by cAMP (ch. 8.5.1). ERE (estrogen response element), and GRE (glucocorticoid response element) are identified respectively by estrogen and glucocorticoid receptors. Estrogen and glucocorticoid are steroid hormones. Although they are not transcription factors, they may affect gene expression through their receptors (ch. 8.1).

Table 5.2 Examples of eukaryotic response elements and transcription factors that bind to them

Name of the sequence	Transcription factor	Consensus sequence
CRE	cAMP response element binding protein (CREB)	TGACGTCA
ERE	Estrogen receptor (ER)	AGGTCANNNTGACCT
GRE	Glucocorticoid receptor (GR)	AGAACANNNTGTTCT
SRE	Serum response factor (SRF)	CC(A/T)$_6$GG
HSE	Heat shock factor 1 (HSF1)	NTTCNNGAANNTTCN

(A/T)6 = six A or T; N = any base.

The serum response factor (SRF) activated by growth factors present in serum binds to the serum response element (SRE). SREs are usually present in the promoters of genes involved in cell cycle progression.

In response to cellular stress (e.g. elevated temperature) heat shock factor (HSF1) is activated. It binds to the HSE (heat shock element) sequences and stimulates the expression of heat shock proteins (HSPs), which enables cell to survive in adverse conditions.

5.4 Structural Motifs of Transcription Factors

Transcription factors must have an ability to interact with DNA. This could be done using several structural protein motifs found in the DNA binding domains (DBDs): zinc finger, leucine zipper, helix-turn-helix, or helix-loop-helix.

The zinc finger motif is elongated protein subunit consisting of 30 amino acids. The structure is stabilized by interactions between the zinc ion and two cysteines and two histidines from side chains. Arranged tandemly zinc fingers recognize longer DNA sequences. One of the first described protein containing this motif was a transcription factor TFIIIA. To transcription factors possessing zinc finger motif belong also: SP1 (binds to DNA in regions rich in GC pairs), estrogen receptor (binds to ERE sequence – estrogen response element), and others steroid hormone receptors. The name of many proteins containing zinc finger motif starts with "ZFP" (from "zinc finger protein").

Leucine zipper binds to DNA as a dimer (homodimer or heterodimer), just like many other transcription factors. It is composed of two chains of α-helix structure, twisted and held together by the hydrophobic interactions between lysine residues. Because lysines are located every seventh amino acid residue, all of them are situated on one side of each helix. Leucine zipper structure recalls the Y letter. Lysines are present in the folded part (from the carboxy terminus), while "arms" (from the amino terminus) are formed from basic amino acids. Such structure allows binding to DNA. The arms can be folded, allowing them to fit into the major groove of DNA. Structures of leucine zipper have been found for example in AP-1 (activator protein 1; heterodimer formed by JUN and FOS proteins) and CREB (ch. 8.5.1).

Helix-turn-helix is a structure containing two α-helices and a short extended chain of amino acids between them. Helix closer to the C-terminus of the protein binds to the major groove of DNA. This motif is present in thousands of proteins binding to DNA. A special variation of the motif is the homeodomain, which is encoded by homeobox genes, playing a key role in the development of the organism.

Helix-loop-helix is formed by two α-helices connected by a loop. Structurally the motif is completely different from motif helix-turn-helix – it is more like leucine zipper. Transcription factors containing helix-loop-helix domain normally function as heterodimers. An example of a transcription factor with such a structure is MyoD (myoblast determination protein).

5.5   Transcription of Protein-Coding Genes

Eukaryotic RNA polymerase II does not have a subunit similar to the prokaryotic σ subunit, which recognizes the promoter and unwinds the DNA double helix. In eukaryotes these functions are performed by proteins called **general transcription factors** (GTFs). RNA polymerase II can interact with six of such factors, marked as TFIIA, TFIIB, TFIID, TFIIE, TFIIF, and TFIIH (TF - transcription factor, II - for the RNA polymerase II). Most of them have a complex structure composed of multiple subunits. The exact order of general transcription factors attachment to the promoter is not known, especially as it can be different for different promoters. In some cases, **preinitiation complex**, containing the general transcription factors and the polymerase, is made independently of the DNA and then it binds to DNA. But usually the process of mounting of the complex to the DNA occurs in a few steps (Fig. 5.5).

In the first step of transcription initiation, **TFIID** factor associates with TATA box sequence in the core promoter. TFIID is composed of two subunits: **TBP** (TATA-box binding protein) and **TAF** (TBP-associated factor). TAF assists TBP in binding to DNA. Some promoters do not have TATA sequence. In this case, TAF binds to DNA first, and then forces TBP binding. In human cells TAF is composed of 12 subunits. One of them can catalyse histones acetylation, which leads to a loosening of interactions between histones and DNA within a nucleosome (nucleosome can be moved and the DNA sequence is accessible for TFs).

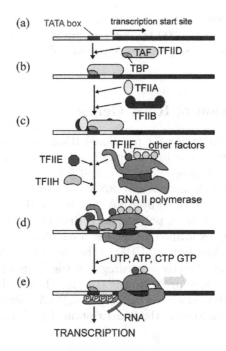

Fig. 5.5 Initiation of transcription in eukaryotes: (a) in most promoters TATA box sequence is present, located about 25 bp upstream the transcription start site; (b) TFIID binds at the place of TATA box; (c) connection of TFIIA stabilizes the complex. TFIIB binds to DNA between TFIID and the future location of the RNA polymerase II; (d) remaining general transcription factors and RNA polymerase II bind to the promoter; (e) TFIIH, using energy from ATP, unwinds DNA at the place of transcription initiation site, which allows starting of RNA synthesis. TFIIH also phosphorylates the polymerase, which leads to the general transcription factors releasing and the beginning of elongation.

After assembly of the preinitiation complex, TFIIH factor unwinds the DNA starting at the position of about -10. Transcription bubble is formed and RNA polymerase II begins synthesis. However, usually only a low (a basal) rate of transcription is driven by the preinitiation complex alone. For modulation of the transcription rate activators and repressors (along with any associated coactivators or corepressors) are responsible (ch. 5.7). During RNA elongation, TFIIF remains associated with the polymerase, while other transcription factors dissociates. Polymerase must be phosphorylated during the elongation.

Elongation of RNA chain proceeds as long as the polymerase encounters a "stop" signal for the transcription. Synthesis of some eukaryotic RNAs is finished in a manner similar to transcription termination in prokaryotes, that is by creating a hairpin structure (containing arm and loop), followed by the U bases. But ussually eukaryotes use different signals to terminate transcription than prokaryotes. The termination mechanism is not exactly known.

Eukaryotic protein-coding genes contain a polyadenylation signal, and often transcription ends in 500-2,000 bp downstream of this signal. Finally, during the RNA processing, sequence downstream of polyadentlation signal is removed (ch. 5.8.2).

5.6   Transcription of RNA Genes

Transcription of genes coding for functional RNA molecules (rRNA, tRNA and small RNA) seems to be less complex than protein-coding genes. In mammals three the rRNA genes (28S, 18S, and 5.8S) are grouped together as a pre-rRNA gene. It is transcribed by RNA polymerase I. tRNA, 5S rRNA, and U6 snRNA are transcribed by RNA polymerase III, while most snRNA genes are transcribed by RNA polymerase II (just like protein-coding genes).

Preinitiation complex formed during the pre-rRNA synthesis is composed of two upstream binding factors (UBFs). They bind to DNA forming a loop between the sequence directly preceding the transcription start site (core element) and upstream control element (UCE). Then TBP (TATA-binding protein), TAFIs (TBP-associated factors), and RNA polymerase I enter the complex, which allows to start the transcription (Fig. 5.6).

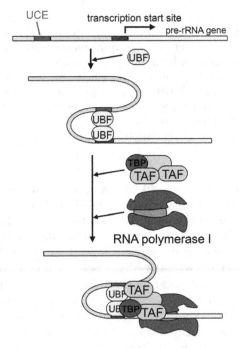

Fig. 5.6 Mounting of the preinitiation complex for RNA polymerase I

Genes encoding tRNA have two regulatory elements (A box and B box) located within the transcription unit. Mounting of the preinitiation complex begins with binding of TFIIIC to these elements. Transcription of the gene encoding the 5S rRNA also depends on regulatory element located within the transcription unit (C box). First TFIIIA binds to it, than, TFIIIC. Joining of further proteins, necessary to start the synthesis of tRNA and 5S rRNA, is carried out analogically: TBP protein enters the complex, then BRF, B" and RNA polymerase III (Fig. 5.7). TBP protein (TATA-binding protein) is necessary to create a preinitiation complex for each of three RNA polymerases.

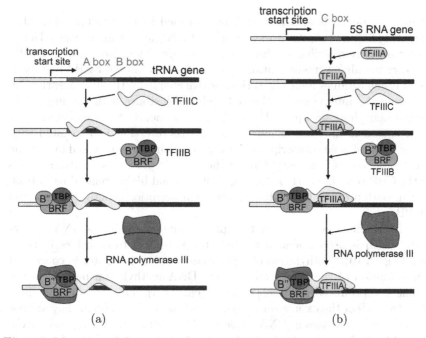

Fig. 5.7 Mounting of the preinitiation complex for RNA polymerase III: (a) tRNA gene transcription; (b) 5S rRNA gene transcription

5.7 The Role of Transcription Factors and Chromatin Structure in Regulation of Transcription

Assembly of the preinitiation complex from the general transcription factors and RNA polymerase II is usually not enough to start the transcription process. With proteins gathered around the transcription start site, another transcription factors (activators) interact. They could be bound to a cis-regulatory elements located far away from the core promoter (Fig. 5.8). RNA molecules are also involved in the regulation of many genes expression.

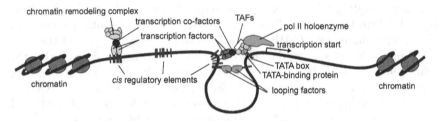

Fig. 5.8 The interactions between transcription factors and changes in chromatin structure during activation of transcription

Gene expression is also regulated by so called looping factors and other interacting factors (coactivators), such as histone acetyltransferases (HATs). Their task is remodeling of chromatin structure. Their presence is usually necessary to enhance transcription by activators. Catalyzed by HAT acetylation of lysine residues neutralizes the positive charge on the histones, thereby decreasing the interaction of the N termini of histones with the negatively charged phosphate groups of DNA. As a consequence, the condensed chromatin is transformed into a more relaxed structure that is associated with greater levels of gene transcription. This relaxation can be reversed by histone deacetylase (HDAC) activity. One of the most important acetyltransferases are the CBP protein (CREB Binding Protein), and highly related to it, p300 protein. Both proteins form a family of the transcription coactivators denoted as CPB/p300.

An additional mechanism that regulates gene expression is DNA methylation, which is attachment of methyl groups ($-CH_3$) to cytosines in DNA chain (Fig. 2.10). Methylation of CpG islands in the promoter is connected with inhibition of the gene transcription. DNA methylation pattern in the genome is reproduced after cell division (ch. 4.5.4). Thus, changes in chromatin that affect the expression of genes are inherited, although they are not associated with changes in DNA sequence. The science describing such mechanisms is called epigenetics. In addition to DNA methylation, to elements of epigenetic gene regulation also belong: modifications of core histones (phosphorylation, methylation, acetylation), the phenomenon of RNA interference (RNAi) (ch. 3.3.4), and gene bookmarking.

5.7.1　Mechanism of Gene Bookmarking

Gene bookmarking is a mechanism of epigenetic memory. It enables to transfer information to the daughter cells which genes have to be active and which might be activated. This results in maintaining the phenotype (i.e. the characteristic features) of a cell line. To create a gene bookmarks on the entire pool of active genes, TFIID and TFIIB general transcription factors (that form a basic preinitiation complex) most frequently are used. During mitosis (cell division), they remain associated with the promoters of genes that were active before the beginning of mitosis. At the same time core histones are modified (Fig. 5.9). This prevents from the complete "packing" of the DNA at these loci during mitosis. Just after mitosis (in early G_1 phase), binding of protein complex that triggers transcription is facilitated in marked loci. Thus, the gene expression profile in cell remains unchanged.

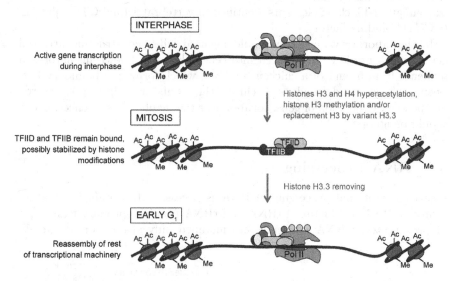

Fig. 5.9 Model of creating gene bookmarks on the entire pool of transcriptionally active genes. Altered from: Sarge K. D. and Park-Sarge O. K. (2005) Gene bookmarking: keeping the pages open. Trends Biochem. Sci. 30, 605-610.

More specific bookmarks are created from proteins of Polycomb (PcG) and Trithorax (TrxG) group, which regulate the expression of homeobox genes, relevant for development of the organism. They interact with DNA within so called cellular memory modules. Also HSF2 transcription factor remains bound with promoter of *Hsp70* genes during the cell division, which keeps them in standby to start transcription immediately after the division, if necessary.

5.7.2 Insulators

Eukaryotic transcriptional activators (enhancers) are able to function over long distances and in an orientation independent manner. Gene expression can be dependent also on chromatin structure. Both enhancers and condensed chromatin could encroach on adjacent genomic domains to perturb gene expression. A class of DNA sequence elements that possess an ability to protect genes from inappropriate signals emanating from their surrounding environment are called insulators. They can block the action of a distal enhancer on a promoter (if the insulator is situated between the enhancer and the promoter). Insulators protect genes also by acting as "barriers" that prevent the advance of nearby condensed chromatin that might otherwise silence expression. Some insulators are able to act both as enhancer blockers and barriers. Others, particularly in yeast, serve primarily as barriers. All characterized enhancer-blocking elements identified in vertebrates bind CTCF protein (CCCTC-binding factor).

In the laboratory, insulators could be used when constructing artificial genes, which are later introduced into cells. The most known insulator is a sequence from 5' end of a chicken β-globin, which forms the boundary between the "open" and "condensed" chromatin. Insulators help to preserve the independent expression of genes located near the regulatory sequences, which should be ignored.

5.8 RNA Processing

Post-transcriptional processing of RNA is necessary to obtain the mature form of mRNA or functional tRNA and rRNA from the primary transcript. In prokaryotes, mRNA molecules are modified only slightly or not at all.

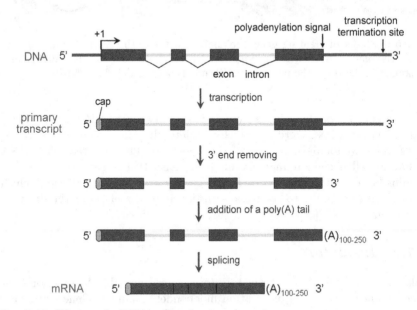

Fig. 5.10 Scheme of mRNA processing in eukaryotes

Many of them undergo translation before the end of transcription. Whereas in eukaryotes, pre-mRNA must undergo post-transcriptional modifications before it can be used in translation. Otherwise mRNA would be destroyed in the cytosol by proteins which task is to destroy the nucleic acids. It is a method of defense against intrusion into the cell of foreign nucleic acid, such as viral.

Pre-mRNA processing includes the following steps (Fig. 5.10):

- capping - addition of guanylate cap, that is 7-methylguanosine (m7G), to the 5' end;
- polyadenylation - addition of the poly(A) tail to the 3' end;
- splicing - removal of introns and connection of exons.

In some cases editing of RNA also occurs.

5.8.1 Capping Process

Addition of a structure called a cap occurs shortly after the start of transcription. The first phosphate group at the 5' end of the transcript is removed by hydrolysis and in its place GTP (guanosine triphosphate) is added losing two phosphate groups: a unique 5'-5' triphosphate linkage is created. Then 7-nitrogen of guanine (N-7) is methylated. In most organisms, the first (and often the second) nucleotide is also methylated at the 2'-hydroxyl of the ribose (Fig. 5.11). Cap affects mRNA stability by protecting the 5' end from phosphatases and nucleases. It is also an important element in the transport of mRNA to the cytoplasm: it interacts with proteins of the exporting complex TREX (transcription export complex protein), which carries a transcript by nuclear pores and brings it to ribosomes.

5.8.2 Polyadenylation

To the 3' end of the transcript adenylate residues are added. In the first stage of polyadenylation, the primary transcript is digested by specific endonuclease recognizing the sequence AAUAAA, which is polyadenylation signal. Cutting occurs about 10 - 35 nucleotides below the sequence. A sequence rich in GU (or U), located about 50 nt below, is an additional determinant of the cleavage site. Next, poly(A) polymerase adds about 250 A residues (in yeast - about 100) to the molecule of RNA (Fig. 5.12). Thanks to poly(A) tail mRNA molecule is recognized as own, not foreign nucleic acid, and is protected against digestion by nucleases. Some viruses, such as influenza virus, can attach to their mRNA poly(A) tail, and by that they are not destroyed in the cytosol. Poly(A) tail also increases the effectiveness of translation.

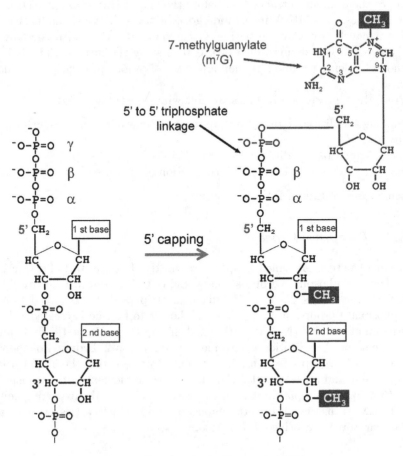

Fig. 5.11 Cap formation on the 5' end of eukaryotic mRNA

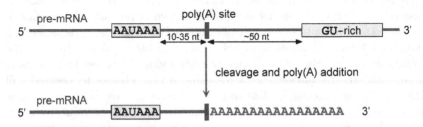

Fig. 5.12 Scheme of polyadenylation of primary transcript in eukaryotes

5.8.3 Processing of Pre-rRNA and Pre-tRNA

In mammals, the pre-rRNA transcript contains three rRNA: 18S, 5.8S and 28S. Transcription and processing occur in the nucleolus. U3 snRNA and other U-rich snRNAs and proteins associated with them are involved in the cutting of pre-rRNA. Synthesized in nucleoplasm 5S rRNA does not require post-transcriptional processing. When it is finished, it moves to the nucleolus and together with 28S rRNA and 5.8 S rRNA forms a large subunit of ribosome.

In eukaryotes, tRNA precursors develop into mature molecules in a series of changes involving:

- removal of the leader sequence (approximately 16 nt) from the 5' end by RNase P;
- removal of intron (approximately 14 nt) from the anticodon loop;
- replacement of two U residues at 3' end of the molecule by CCA;
- modification of some bases.

5.8.4 RNA Splicing

In the process of splicing, introns are removed from primary transcript and exons are joined together. Splicing must occur very precisely, because even a single-nucleotide shift can move the reading frame and completely change a sequence of amino acids. Well-defined signal for splicing is usually present in introns. Most introns begin with GU sequence from the 5' end (splice donor site) and end with AG (splice acceptor site). Also the presence of the CU(A/G)A(C/U) sequence is important (where A is preserved in all genes). It is called a branch site and is located 20-50 nt upstream from the splice acceptor site (Fig. 5.13). In 60% of cases exon has the sequence (A/C)AG in the donor site, and G in the acceptor site. Moreover, auxiliary cis-elements, known as exonic and intronic splicing enhancers (ESEs and ISEs) and exonic and intronic splicing silencers (ESSs and ISSs), aid in the recognition of exons.

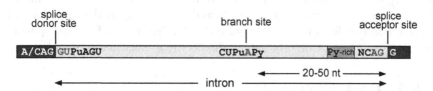

Fig. 5.13 The classical splicing signals found in the major class (>99%) of human pre-mRNAs. Pu — purine (A or G), Py — pyrimidine (C or U). Py-rich — sequence rich in pyrimidine. Auxiliary elements are not shown.

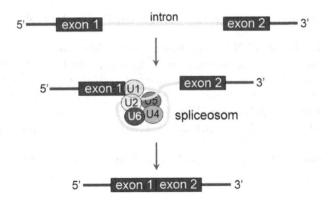

Fig. 5.14 Scheme of spliceosome formed during intron removing. U1-U6 — snRNA-protein complexes (snRNP).

Components of the basal splicing machinery bind to the classical splice-site sequences and promote assembly of the multicomponent splicing complex known as the spliceosome (Fig. 5.14). The spliceosome performs the two primary functions of splicing: recognition of the intron/exon boundaries and catalysis of the cut-and-paste reactions that remove introns and join exons. The spliceosome is made up of five small nuclear ribonucleoproteins (snRNPs) and more than one hundred proteins. Each snRNP is composed of a single uridine-rich small nuclear RNA (snRNA) and multiple proteins. The U1 snRNP binds the 5' splice site, and the U2 snRNP binds the branch site via RNA:RNA interactions between the snRNA and the pre-mRNA. Spliceosome assembly is highly dynamic. Both splice sites are recognized multiple times by interactions with different components during the course of spliceosome assembly.

The typical human gene contains an average of 8 exons. Internal exons average 145 nucleotides (nt) in length, and introns average more than 10 times this size and can be much larger. Most pre-mRNAs undergo alternative splicing. In such cases some splicing signals are masked by the regulatory proteins. Alternative splicing is a major mechanism for modulating the gene expression of an organism. It enables a single gene to increase its coding capacity, allowing the synthesis of several structurally and functionally distinct mRNA and protein isoforms from a unique gene. However, disruption of normal splicing patterns can cause or modify human diseases (e.g. muscular dystrophy).

5.8.5 RNA Editing

Information contained in some RNA molecules can be changed just after the end of the transcription in a process different than splicing. Change of a single nucleotide in mRNA may cause a codon change. For example deamination of cytosine in glutamine codon CAA leads to the formation of uracil and UAA stop codon, and so, to earlier termination of translation as it is in *apo-B* gene (Fig. 5.15). In mammals *apo-B* gene is expressed in liver and in epithelium of small intestine. In liver, 500 kDa protein is created (called Apo-B100), in intestine — much smaller (called Apo-B48). Apo-B100 is produced without RNA editing, while Apo-B48 is produced from mRNA, in which cytosine in CAA codon has been deaminated by an enzyme present in epithelial cells of intestine (but not in liver).

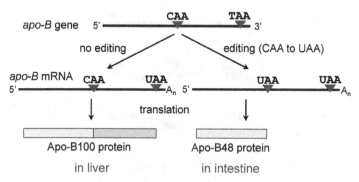

Fig. 5.15 Editing of the apolipoprotein B RNA

5.9 Reverse Transcription

In RNA viruses (called retroviruses), RNA is the carrier of genetic information. After entering a host cell, viral RNA is transcribed to a double-stranded DNA (in a process called reverse transcription) and is integrated into the host DNA. Reverse transcription is catalyzed by reverse transcriptase, an enzyme which is introduced into the host cell by virus during infection. Reverse transcriptase catalyzes three types of reaction: synthesis of DNA on RNA template, synthesis of DNA on DNA template, and RNA hydrolysis (Fig. 5.16).

Transcription of retroviral DNA occurs only when it is integrated into the DNA of the host (Fig. 5.17). Integration occurs thanks to IN protein (integration protein; integrase) which is introduced into the cell by the virus together with the reverse transcriptase. IN is an endonuclease, which recognizes specific sequences in LTR. Integration of viral DNA can occur in any place in the host genome.

Genetic strategy used by retroviruses can be observed in various eukaryotic genetic elements, for example in Ty elements found in yeast (T – transposon, y – yeast). Ty1 is a transposon (i.e. mobile genetic element) of length of 6 kbp, integrated with the yeast genome. It has a border LTR sequences and is flanked by short repeated sequences. Ty1 encodes proteins similar to the

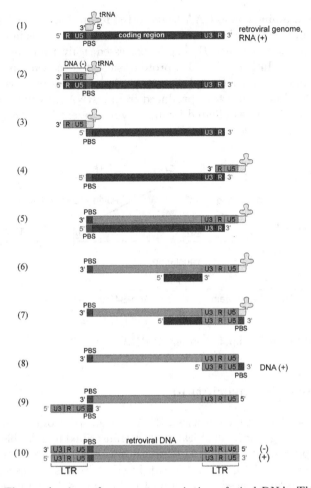

Fig. 5.16 The mechanism of reverse transcription of viral RNA. The process is catalysed by reverse transcriptase — enzyme that has DNA polymerase and RNase H activities. (1) Cellular tRNA specific for the retrovirus hybridizes with a complementary region of the virus genome called the primer binding site (PBS). (2) DNA synthesis begins from the 3'-OH group of attached tRNA on the template of retrovirus genomic RNA. DNA fragment (-) complementary to the sequence of U5 and R is created. (3) The viral U5 and R sequences are removed by the enzyme domain with RNase H activity. (4) First shift: DNA strand (-) hybridizes with the R sequence present in the 3' end of viral RNA. (5) DNA strand (-) is extended from the 3' end. (6) Most viral RNA is removed by RNase H. (7) The remaining fragment of viral RNA is used as primer for synthesis of the second strand of DNA (+). (8) The remaining viral RNA and tRNA are removed by RNase H. (9) Second shift: PBS region of the second DNA strand (+) hybridizes with the PBS region of the first DNA strand (-). (10) Synthesis of the missing parts of both strands. At the ends of retroviral DNA, long terminal repeats, called LTR sequences, are formed. U3 and U5 have different unique sequences, R is a repeated sequence. LTRs contain signal sequences for transcription and integration processes.

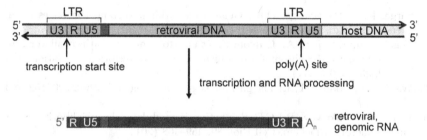

Fig. 5.17 Formation of retroviral genomic RNA

retroviral reverse transcriptase and integrase. Duplication and transfer (transposition) to any other place in the genome occurs with the participation of an intermediate product — RNA, which is produced in transcription process. Then, on Ty RNA template, double-stranded DNA is synthesized, which is included to the genome. So, the flow of Ty transposon information (from DNA to RNA and back to the DNA) is the same as in retroviruses. Genetic elements that use this strategy are called retrotransposons or retroposons (ch. 7.4.3). Similar elements were found in Drosophila (copia elements).

References

1. Berg, J.M., Tymoczko, J.L., Stryer, L.: RNA Synthesis and Splicing. In: Biochemistry, 5th edn. W. H. Freeman, New York (2002)
2. Clancy, S.: DNA transcription. Nature Education 1(1) (2008), http://www.nature.com/scitable/topicpage/dna-transcription-426
3. Clancy, S.: RNA splicing: introns, exons and spliceosome. Nature Education 1(1) (2008), http://www.nature.com/scitable/topicpage/rna-splicing-introns-exons-and-spliceosome-12375
4. Hoopes, L. (ed.): Gene Expression and Regulation. In: Miko, I. (ed.) Genetics. Nature Education (2011), http://www.nature.com/scitable/topic/gene-expression-and-regulation-15
5. Katahira, J., Yoneda, Y.: Roles of the TREX complex in nuclear export of mRNA. RNA Biol. 6, 149–152 (2009)
6. King, M.W.: Regulation of Gene Expression. In: themedicalbiochemistrypage.org, LLC (1996-2012), http://themedicalbiochemistrypage.org/gene-regulation.php
7. King, M.W.: RNA Transcription & Processing. In: themedicalbiochemistrypage.org, LLC (1996-2012), http://themedicalbiochemistrypage.org/rna.php
8. Lodish, H., Berk, A., Zipursky, S.L., et al.: Regulation of Transcription Initiation. In: Molecular Cell Biology, 4th edn. W.H. Freeman, New York (2000)
9. Lodish, H., Berk, A., Zipursky, S.L., et al.: RNA Processing, Nuclear Transport, and Post-Transcriptional Control. In: Molecular Cell Biology, 4th edn. W.H. Freeman, New York (2000)
10. Proudfoot, N.J.: Ending the message: poly(A) signals then and now. Genes Dev. 25, 1770–1782 (2001)

11. Sarge, K.D., Park-Sarge, O.K.: Gene bookmarking: keeping the pages open. Trends Biochem. Sci. 30, 605–610 (2005)
12. Vogelmann, J., Valeri, A., Guillou, E., et al.: Roles of chromatin insulator proteins in higher-order chromatin organization and transcription regulation. Nucleus 2, 358–369 (2011)
13. Ward, A.J., Cooper, T.A.: The pathobiology of splicing. J. Pathol. 220, 152–163 (2010)
14. Weth, O., Renkawitz, R.: CTCF function is modulated by neighboring DNA binding factors. Biochem. Cell Biol. 89, 459–468 (2011)
15. Yang, J., Corces, V.G.: Chromatin insulators: a role in nuclear organization and gene expression. Adv. Cancer Res. 110, 43–76 (2011)

Chapter 6
Synthesis and Posttranslational Modifications of Proteins

6.1 Protein Synthesis

Protein synthesis (translation) is a process in which more than one hundred macromolecules work together in a coordinated way. Translation is carried out on ribosomes by tRNA, and on the mRNA template sequence (ch. 3.3). Messenger RNA sequence is translated to the protein sequence. Translation starts from the 5' end of the mRNA and follows to 3' end strictly in accordance with ternary code (three nucleotides for one amino acid), that is according to the reading frame. Even a small change in the mRNA sequence can lead to the production of completely different protein, for example when amino acid codon is changed to a stop codon produced protein will be shorter. Frameshift can also occur.

A peptide chain is synthesized in the direction from amino terminus to carboxyl terminus. It always starts from methionine, encoded by the **AUG** codon (initiation codon). In bacteria, a single mRNA molecule can encode multiple proteins (it is called a **polycistronic transcript**), so it contains several initiation codons (Fig. 6.1). A protein may also contain several inner methionines, also encoded by the AUG codon. AUG codons that serve as initiation codons must be somehow distinguished from inner AUG codons. This distinction occurs thanks to so called **initiation signals**.

6.1.1 The Initiation Signal for Protein Synthesis

The whole process of translation is very similar in prokaryotic and eukaryotic cells. However, they have different mechanisms to recognize the initiation codon. Many mRNA in prokaryotes are polycistronic, that means they encode more than one protein. Such mRNA contains a number of initiation codons. They are recognized thanks to the specific sequence (UAAGGAGG)

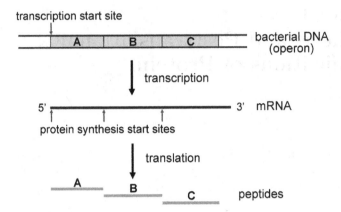

Fig. 6.1 Expression of a bacterial operon, i.e. DNA fragment encoding several transcribed together proteins, for example structural proteins, preceded by common control sequences — promoter and operator

located 5-10 nt upstream of initiation codon. It is called **Shine-Dalgarno** sequence (named after its discoverers). Ribosomal 16S rRNA molecule contains a sequence complementary to the Shine-Dalgarno sequence:

```
5'- UAAGGAGG(5-10 nt)AUG     mRNA
3'- AUUCCUCC........         16S rRNA
```

Association of these two sequences leads to joining of the remaining parts of ribosome.

Mechanism of initiation codon recognition in eukaryotes is not quite clear. Perhaps eukaryotic ribosome simply scans the mRNA starting from the cap at the 5' end in search of the first AUG codon. It seems possible since almost all eukaryotic mRNA are monocistronic (encode a single protein). However some viral mRNA are polycistronic or do not have a cap, and are still translated in eukaryotic cell. It has been proved, that so called **Kozak** sequence (named after its discoverer) — 5'-ACCAUGG — located in mRNA around the initiation codon, increases the efficiency of translation. Also communication between the 5' cap and the 3' poly(A) tail of mRNA via specific proteins (that leads to mRNA circularization) results in the enhancement of translation.

6.1.2 Role of tRNA in Translation

Amino acids alone are unable to recognize codons in the mRNA. The amino acid needs to be delivered to the ribosome in the form associated with the specific tRNA — such one, which anticodon matches the amino acid codon.

Molecule of tRNA with amino acid attached at the 3' end is called the **aminoacyl-tRNA** (or charged tRNA). Attachment of amino acid to tRNA is catalyzed by different aminoacyl-tRNA synthetases (also known as activating enzymes). They recognize anticodon and the corresponding amino acid. There are 20 different aminoacyl-tRNA synthetases present in the cell. Each can attach only one from 20 amino acids to the corresponding tRNA. Notation "aa-tRNA" (amino acid-tRNA, e.g. Lys-tRNA) stands for tRNA with attached amino acid (e.g. lysine), while notation "tRNA^{aa}" (e.g. tRNA^{Lys}) stands for tRNA with anticodon for the codon of given amino acid. Notation "Lys-tRNA^{Lys}" represents the specific for lysine tRNA with lysine attached.

Once aminoacyl-tRNA enters the ribosome, complementation between its anticodon and corresponding codon in mRNA takes place. An amino acid is added to synthesized peptide (actually — synthesized peptide to amino acid). In bacteria, there are 30 to 40 tRNA molecules, each with different anticodon. In higher organisms, there are about fifty different tRNAs. What's more there are 61 different codons for amino acids, that is much more than tRNA. Assuming that each codon will create pair only with completely matching anticodon, 61 different tRNA would have to exist.

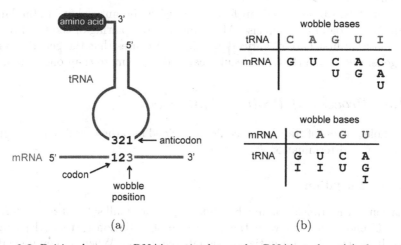

Fig. 6.2 Pairing between tRNA's anticodon and mRNA's codon: (a) the spatial criteria for the pairing of third base from codon are less restrictive than in the case of other two bases; (b) possibilities of base pairing at the third codon position: for example G can form a pair with C or U, whereas I — with C, A or U

The disproportion between the number of codons and the number of tRNA molecules results from the fact, that there is some spatial tolerance in pair creation of the third base in codon, which is called wobble position (Fig. 6.2a). At this position, the normal rules for pairing are not valid: one base can create a pair with one of several bases. Frequently, the first anticodon base (which pairs with base at third position of the codon) is inosine (I). This allows

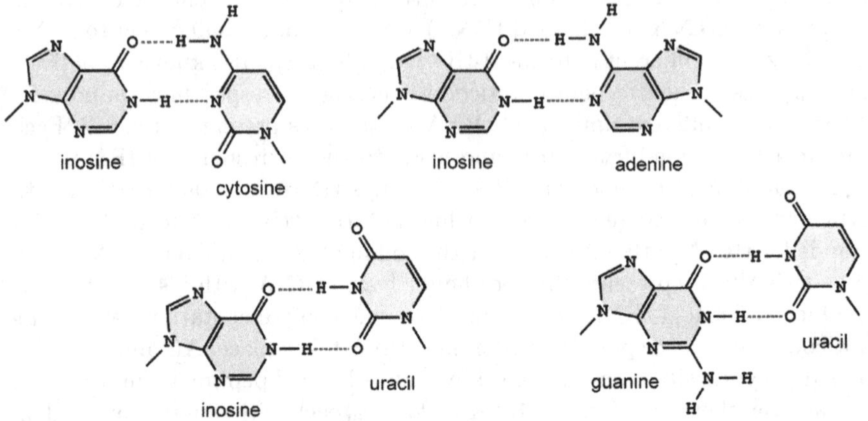

Fig. 6.3 Wobble base pairs for inosine and guanine

tRNA molecule reading of the maximum number of three codons (Fig. 6.2b and 6.3).

The first two bases in codon form pairs with the anticodon in standard manner — recognition is precise. More unrestricted pairing of the third base in codon with first base of anticodon (Fig. 6.3) causes that the genetic code is degenerated: some tRNA molecules can recognize more than one codon.

6.1.3   Process of Protein Synthesis

The whole process of translation could be divided into three stages: initiation, elongation and termination.

6.1.3.1   Initiation

Initiation of the translation involves binding of the small subunit of ribosome to the 5' end of mRNA with the assistance of initiation factors (IF) (Fig. 6.4a). It is dependent on initiation signals (ch. 6.1.1). When all components of the translational machinery are recruited, a peptide elongation starts.

Because protein synthesis always begins with methionine, the first aminoacyl-tRNA is Met-tRNA$_i^{Met}$, where "i" stands for the initiation (although methionine is not always the first amino acid of functional protein: N-terminus could be cut off after synthesis). In bacteria, the methionine in initiator aminoacyl-tRNA is modified by addition of the formyl group (-HCO) to the amino group. This creates a **formylmethionine** (fMet), which is unique for bacteria. In mammals, it induces a strong immune response.

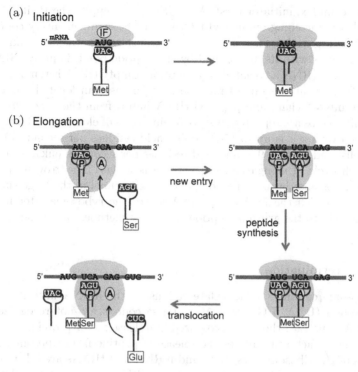

Fig. 6.4 Stages of protein synthesis: (a) without mRNA, large and small ribosome subunits exist separately. Protein synthesis begins when the initiation factors (IF) form complex with the small subunit of ribosome, mRNA and the initiator aminoacyl-tRNA. Next, the large subunit of ribosome joins, and initiation factors are released. (b) In process of elongation one cycle comprises the following steps:
— new entry: new aminoacyl-tRNA with appropriate anticodon is bound in the A site. In prokaryotes this phase is catalyzed by elongation factors Tu and Ts, in eukaryotes — by EF_1 and $EF_{1\beta}$.
— peptide synthesis: the growing polypeptide connected to the tRNA in the P site is transferred to the aminoacyl-tRNA present in A site. This process is known as peptide bond formation. Now, the A site has the newly formed peptide, while the P site has an uncharged tRNA (tRNA with no amino acids).
— translocation: the ribosome moves 3 nucleotides toward the 3' end of the mRNA. Since tRNAs are linked to mRNA by codon-anticodon base-pairing, the tRNAs move relative to the ribosome, taking the nascent polypeptide from the A site to the P site and moving the uncharged tRNA to the exit from the ribosome. In bacteria translocation is catalyzed by elongation factor G, in eukaryotes — the elongation factor EF_2

6.1.3.2   Elongation

A ribosome contains two main tRNA binding sites: **A site** (aminoacyl) and **P site** (peptidyl). After binding of a large subunit of ribosome with the

initiation complex, initiator (first) Met-tRNA$_i^{Met}$ occupies the P site. The A
site binds the incoming aminoacyl-tRNA with the complementary codon on
the mRNA (Fig. 6.4b). Then methionine is transferred to the new aminoacyl-
tRNA and joins with its amino acid by the peptide bond. Peptidyl-tRNA is
formed, that is tRNA associated with peptide. Empty tRNA is removed from
the P site, and, at the same time, the ribosome is translocated one codon
down the mRNA chain and peptidyl-tRNA jumps from the A site to the P
site. Similar steps are repeated in subsequent cycles of elongation. The rate of
translation varies; it is up to 17-21 amino acid residues per second in prokaryotic cells, and up to 6-9 amino acid residues per second in eukaryotic cells.
After each cycle of elongation, the peptide is longer. The growing protein
exits the ribosome through the polypeptide exit tunnel in the large subunit.
The site by which the uncharged tRNA leaves the rybosome after it gives
its amino acid to the growing peptide chain is sometimes called the exit site
(E site).

6.1.3.3 Termination

The ribosome continues to translate codons on the mRNA until it reaches
a stop codon (UAA, UGA or UAG). These codons are not recognized by
any tRNAs. Instead, they are recognized by release factor proteins. Association of release factor stimulates the release of peptide from ribosome. Next,
subunits of the ribosome dissociate and mRNA and tRNAs are released. All
translational components are now free for additional rounds of translation.

6.1.4 Reading Frame Shift

Reading frame shift (Fig. 6.5) is caused in most cases by insertions or deletions
of a single nucleotide in mRNA (ch. 4.4). In such situation a completely
different protein is produced.

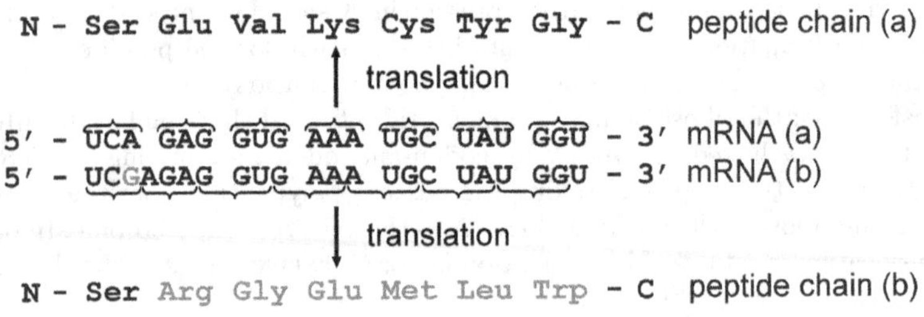

Fig. 6.5 An example of reading frame shift. mRNA(a) and mRNA(b) differ only in
one nucleotide — it is an additional G nucleotide at the third position of mRNA(b)
(marked in green). The amino acid sequence synthesized on mRNA(b) template is
completely different starting from the insertion site.

6.1.5 Overlapping Genes

Overlapping genes are known to be common in compact genomes (viruses, bacteria, mitochondria). Increasing evidence suggests the existence of many overlapping genes also in eukaryotic genomes. In the same-strand overlapping type the shifting of reading frame enables coding for up to three completely different proteins from one strand of DNA sequence. In eukaryotes, such type of the gene overlapping occurs in mitochondrial DNA, where genes of, for example, ATPase subunits 6 and 8 overlap (Fig. 6.6). ATPase 8 is composed of 68 amino acids, ATPase 6 - of 226. Their amino acid sequence is different even in the region encoded by a shared DNA sequence. It is interesting that in the ATPase 6 gene there is no classical stop codon, only the first two nucleotides (TA). Full stop codon (TAA) is formed only after the addition of the poly(A) tail during RNA processing.

Fig. 6.6 An example of the same-strand gene overlapping

Another category of overlapping genes is based on direction of transcription. Transcription from the same sequence in opposite directions (the different-strand overlapping type) leads to generation of antisense transcripts. Their functions in living organisms are not fully understood. They could affect the regulation of gene expression at the level of transcription, mRNA processing, splicing, or translation.

6.2 Posttranslational Modifications and Protein Sorting

A typical mammalian cell contains up to 10,000 different kinds of proteins. For a cell to function properly, each of its numerous proteins must be localized to the correct compartment. A few proteins are synthesized on ribosomes in mitochondria and chloroplasts and are used in those organelles. However, most proteins (even mitochondrial and chloroplast) are synthesized on ribosomes in the cytosol. They must be distributed to their correct destinations. This is possible thanks to the signal sequences present in emerging peptides. In general, segregation pathways could be divided to associated and not associated with the endoplasmic reticulum (ER) (Fig. 6.7). Transport of proteins to nucleus, mitochondria, chloroplasts and peroxisomes is not associated with ER. Those proteins are synthesized on free ribosomes as soluble polypeptides. Proteins directed to peroxisomes usually contain the sequence "Ser-Lys-Leu"

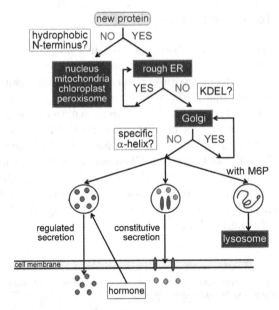

Fig. 6.7 General scheme of proteins sorting:1. If at the N-terminus of the new peptide a fragment with hydrophobic amino acids is present, the peptide is transferred to the endoplasmic reticulum (ER) for further sorting. If there is no hydrophobic N-terminus, sorting is not associated with reticulum. 2. The peptide is retained in the ER if its C-terminus contains the sequence "Lys-Asp-Glu-Leu" (KDEL in one-letter code). If not — it is moved to the Golgi apparatus. 3. In Golgi apparatus, proteins having a specific transmembrane helix remain. 4. After glycosylation at the Golgi apparatus, modified proteins containing a mannose 6-phosphate (M6P) are directed to lysosomes. 5. Proteins that aggregate with chromogranin B (secretogranin I) or secretogranin II are transferred to the secretory vesicles, from which they are released upon specific stimulation (controlled secretion). Other types of vesicles move toward the cell membrane and their content is constantly released outside the cell (constitutive secretion).

(SKL in one-letter code) at the C terminus, while directed to mitochondria — amphipathic helix 20-50 residues with R/K and hydrophobic sides. Transport to the nucleus is completely different — it is described in chapter 6.2.4.

Programs available on the websites can be used to predict the protein subcellular localization on the basis of amino acid sequence:
http://www.geneinfinity.org/sp/sp_proteinloc.html

6.2.1 Targeting of Proteins to the Rough Endoplasmic Reticulum

Secretory proteins and membrane proteins are synthesized on ribosomes associated with the endoplasmic reticulum (so-called rough ER; the presence of these bound ribosomes distinguishes the rough ER from the smooth ER).

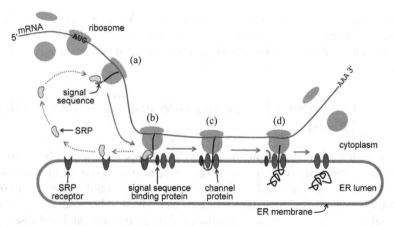

Fig. 6.8 Introduction of newly synthesized peptide into rough ER: (a) after synthesis of about 70 amino acids, signal sequence present at the amino terminus of the polypeptide chain emerges from the ribosome. Signal sequence is recognized and bound by a signal recognition particle (SRP). Translation is inhibited. SRP directs the peptide with the ribosome toward the ER membrane; (b) SRP binds to its receptor on the ER membrane. The ribosome then binds to a protein translocation complex in the ER membrane, while SRP is released; (c) signal sequence is inserted into a membrane channel in assistance of the signal sequence binding protein. Upon docking, polypeptide chain elongation resumes; (d) signal sequence is cleaved and degraded. After the end of protein synthesis the polypeptide is released into the lumen of the ER, ribosomal subunits detach from the mRNA and can re-start searching for template. Released SRP also can bind another, emerging from the ribosome signal sequence.

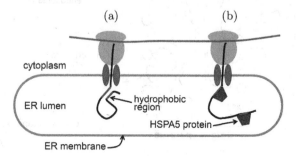

Fig. 6.9 Prevention of abnormal folding of proteins in the ER: (a) peptide may begin folding before the end of synthesis (hydrophobic fragments have a tendency to aggregate); (b) HSPA5 protein binds to the hydrophobic parts of emerging peptide before they are incorrectly folded and prevents premature aggregation

Whether the ribosome will be free or attached to rough ER is determined only by the type of protein synthesized on it. If the protein contains signal sequence including a stretch of hydrophobic residues at the amino terminus, it will be targeted to the ER during their translation (Fig. 6.8).

Inside the ER the process of protein folding takes place. Heat shock proteins, dependent on the ATP, serve as chaperones which bind the growing proteins and help in their folding (Fig. 6.9). Similar mechanisms exist also in cytoplasm.

6.2.2 Transport of Proteins into and across the Golgi Apparatus

Proteins without KDEL sequence at the C-terminus are directed from the endoplasmic reticulum to Golgi apparatus. Transport is mediated by small vesicles. The Golgi is composed of stacks of membrane-bound structures known as cisternae. An individual stack is sometimes called a dictyosome, especially in plant cells. Each cisternae comprises a flat, membrane enclosed disc that includes special Golgi enzymes which modify or help to modify cargo proteins that travel through it (Fig. 6.10).

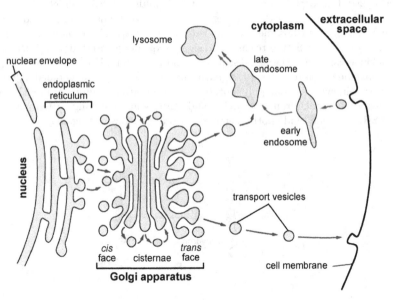

Fig. 6.10 The Golgi apparatus. Transport of proteins into and across the Golgi is mediated by small vesicles

The main functional regions of the Golgi are: the cis-Golgi network (*cis* face), medial-Golgi, endo-Golgi, and trans-Golgi network (*trans* face). Cis face is directed to the endoplasmic reticulum. Vesicles from the ER fuse with the network and subsequently progress through the stack to the *trans* face, where they are packaged and sent to the required destination. Each region contains different enzymes which selectively modify the contents depending on where they reside. *Cis* face is additionally an "emergency compartment" for the proteins produced in the ER, which were accidentally caught in vesicles flowing into Golgi apparatus (they are directed back). Vesicles released from the *trans* face (maturation pole) deliver proteins and lipids to the cell membrane, lysosomes, exosomes, endosomes, and to other destinations.

6.2.3 Transport of Proteins into Lysosomes

After glycosylation at the Golgi apparatus proteins are delivered to lysosomes if one or more molecules of mannose are phosphorylated at the carbon atom at position 6. *Trans* face of Golgi apparatus contains mannose-6-phosphate receptor, which binds to mannose-6-phosphate present on lysosomal enzymes and delivers them to secretory vesicles targeted to lysosomes.

6.2.4 Nuclear Transport

Many proteins produced in the cytoplasm move to the nucleus where they perform their function. On the other hand RNA molecules produced in the nucleus have to be transported to the cytoplasm, where there are necessary for the protein synthesis. Macromolecules require association with karyopherins called **importins** to enter the nucleus, and **exportins** to exit.

Small proteins with mass of 15 kDa, such as histones, can pass through nuclear pores without the assistance of other proteins. Large proteins (above 90 kDa) can not pass through the nuclear membrane if they do not have specific signals. Such signals may also accelerate the entry of small proteins into the nucleus. Specific signal sequence is either directly recognized by importins and exportins or through an adapter protein. For example, importin β does not recognize a specific sequence of transported protein, but it is supported by the importin α being the adapter. On the other hand, importin for hnRNP (heterogeneous nuclear ribonucleoprotein) recognizes a specific sequence of hnRNP by itself. This importin is called transportin.

6.2.4.1 Nuclear Localization Signal and Nuclear Export Signal

Protein that must be imported to the nucleus from the cytoplasm contains nuclear localization signal (NLS) that is recognized by importins. NLS is a sequence of amino acids that acts as a tag. Two types of such signals have been identified: SV40 type and bipartite type. SV40 type signal was found for the first time in the large T antigen of the SV40 virus. It has amino acid sequence: PKKKRKV. It is characterized by the presence of several consecutive basic residues; in many cases it also contains a proline residue.

The bipartite type was first discovered in *Xenopus* neoplasmin. It contains the sequence: **KR[PAATKKAGQA]KKKK**. Characteristic features are: two basic residues, 10 spacer residues, another basic region containing at least 3-5 basic residues.

Nuclear export signal (NES) is a leucine-rich (e.g.: **LQLPPLERLTL**) domain recognized by a class of exportins called exportin 1 or Crm1. Proteins, which are shuttled between nucleus and cytoplasm, contain both transport signals: NLS and NES.

6.2.4.2   Mechanism of Importins and Exportins Action

Exportins and importins action is regulated by a G protein (guanine nucleotide-binding protein) called **Ran**. It belongs to the family of monomeric G proteins (known as small G proteins; there is also heterotrimeric G proteins family — ch. 8.3). Like other G proteins, Ran may be associated with GDP (guanosine diphosphate) or GTP (guanosine triphosphate). Switch from the GTP-bound to the GDP-bound state is catalyzed by a GTPase-activating protein (GAP), which hydrolyzes GTP bound to the protein. The reverse transition is catalyzed by guanine nucleotide exchange factor (GEF), which induces exchange between GDP associated with the protein and the cellular GTP (Fig. 6.11).

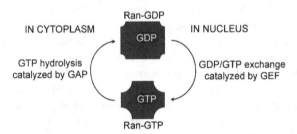

Fig. 6.11 Two forms of Ran: associated with GTP and associated with GDP

The GEF is located predominantly in the nucleus while GAP — almost exclusively in the cytoplasm. Therefore, in the nucleus Ran will be mainly loaded with GTP (Ran-GTP) while cytoplasmic Ran will be mainly in the GDP-bound state (Ran-GDP). Binding of exportin or importin to their cargo is dependent on the interaction with Ran: Ran-GTP enhances interactions between an exportin and its cargo, and stimulates the release of cargo from importin; Ran-GDP stimulates the release of exportin's cargo and enhances the binding between an importin and its cargo. When the complex exportin-cargo-Ran-GTP leaves the nucleus (through nuclear pores), GTP is hydrolyzed by GAP present in the cytoplasm and the cargo is released

(Fig. 6.12). Similarly, importin with its cargo can move in the cytoplasm, but the cargo is released in the nucleus, where Ran-GDP is converted to Ran-GTP by GEF.

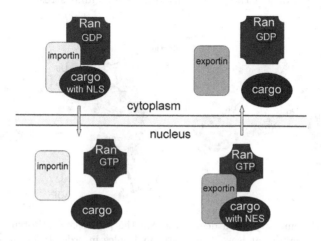

Fig. 6.12 General function of importins and exportins

6.3 Protein Degradation

Some proteins remain in the cell very shortly (e.g. enzymes regulating the metabolism of the cells exist only a few minutes), others are very stable (remain for several days). Proteins that have fulfilled their function in the cell or damaged proteins are targeted for degradation. Controlled protein degradation is essential for the proper function of the cell. It is curried out either in lysosomes (which contain active proteolytic enzymes) or in proteasomes. Proteins taken in by endocytosis (derived from the outside of the cell) and long-living proteins are mainly degraded in lysosomes. They are surrounded by cell membrane, which then is fused with lysosomes. Short-living and damaged proteins are degraded in proteasomes.

Proteasomes (26S protease complex) are large protein complexes present both in cytoplasm and nucleus. Four stacked rings, each composed of seven protein subunits, create core of the proteasome. At each end of the core, regulatory units (caps) are situated. Centrally, inside the rings, catalytic center with proteolytic properties is located, in which proteins are degraded. Openings at the two ends of the core allow the target protein to enter.

Proteins, which are designed for destruction in the proteasome, are marked by attachment of a few **ubiquitin** molecules. Ubiquitin is a small protein with a mass 8.5 kDa. It is present in almost all tissues (ubiquitously) of eukaryotic organisms and is conserved in evolution. A protein marked with ubiquitins is digested by the proteasome while ubiquitins are released and could be used again (Fig. 6.13). Some proteins are degraded independently of ubiquitination. This applies mainly to unstable proteins, which have not been properly folded or lost their structure during a cellular stress.

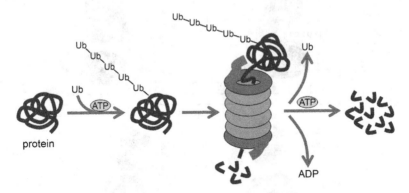

Fig. 6.13 Scheme of protein degradation in the proteasome. Protein must have attached the chain of at least 4 ubiquitin molecules in order to be recognized by a regulatory proteasome subunit (red). Three different enzymes and energy from ATP are necessary for the ubiquitin (Ub) attachment. Before the protein moves into the proteasome core it must be at least partially unfolded, which also needs the energy from ATP. After passing through the central part of the proteasome proteins are usually degraded to 7-9 amino acids fragments.

The proteasomal degradation pathway is essential for many cellular processes. For the discovery of ubiquitin-mediated protein degradation the 2004 Nobel Prize in Chemistry was awarded to Aaron Ciechanover, Avram Hershko and Irwin Rose.

As a result of degradation in the proteasome generally very short peptides are released. Sometimes however, the product of digestion in the proteasome is functional, biologically active molecule. Some transcription factors, such as components of NF-κB, are synthesized as an inactive precursors. Their ubiquitination and degradation in the proteasome is required for transformation into an active form. Such mechanism of selective degradation is known as regulated ubiquitin/proteasome dependent processing (RUP).

References

1. Berg, J.M., Tymoczko, J.L., Stryer, L.: Protein Synthesis. In: Biochemistry, 5th edn. W.H. Freeman, New York (2002)
2. Clancy, S., Brown, W.: Translation: DNA to mRNA to protein. Nature Education 1(1) (2008), http://www.nature.com/scitable/topicpage/translation-dna-to-mrna-to-protein-393

3. Cooper, G.M.: Protein Degradation. In: The Cell: A Molecular Approach, 2nd edn. Sinauer Associates, Sunderland, MA (2000),
   http://www.ncbi.nlm.nih.gov/books/NBK9957

4. King, M.W.: Translation of Proteins. In: themedicalbiochemistrypage.org, LLC (1996-2012),
   http://themedicalbiochemistrypage.org/protein-synthesis.php

5. King, M.W.: Protein Modifications. In: themedicalbiochemistrypage.org, LLC (1996-2012),
   http://themedicalbiochemistrypage.org/protein-modifications.php

6. Lodish, H., Berk, A., Zipursky, S.L., et al.: Protein Sorting: Organelle Biogenesis and Protein Secretion. In: Molecular Cell Biology, 4th edn. W.H. Freeman, New York (2000)

7. Makalowska, I., Lin, C.F., Makalowski, W.: Overlapping genes in vertebrate genomes. Comput. Biol. Chem. 29, 1–12 (2005)

8. Mazumder, B., Seshadri, V., Fox, P.L.: Translational control by the 3'-UTR: the ends specify the means. Trends Biochem. Sci. 28, 91–98 (2003)

9. Muratani, M., Tansey, W.P.: How the ubiquitin-proteasome system controls transcription. Nat. Rev. Mol. Cell. Biol. 4, 192–201 (2003)

10. Poon, I.K., Jans, D.A.: Regulation of nuclear transport: central role in development and transformation? Traffic 6, 173–186 (2005)

11. Rape, M., Jentsch, S.: Productive RUPture: activation of transcription factors by proteasomal processing. Biochim. Biophys. Acta 1695, 209–213 (2004)

Chapter 7
Cell Division

7.1 Cell Cycle

As the cell grows and divides, it progresses through stages in the cell cycle. The eukaryotic cell cycle could be defined as a period from one cell division to the next one. It consists of four phases: G_1, S, G_2 and M, where "G" stands for gap, "S" — synthesis, and "M" — mitosis. During the gaps cell increases in size. After the division, cell may initiate a new round of division or can remain for a longer period in resting phase, G_0. Cell, after appropriate stimulation, can leave the G0 phase and re-enter the G_1 phase (Fig. 7.1). In most mammals the cell cycle lasts 12-24 hours (without G_0 phase). Bacteria divide much faster, for example *E. coli* — every 20-30 minutes (however, there is no typical cell cycle in bacteria).

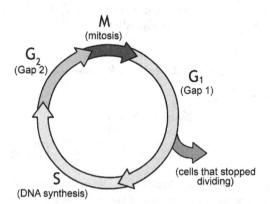

Fig. 7.1 Eukaryotic cell cycle phases

Proper regulation of the cell cycle is crucial to the survival of a cell. Among others, the cell has to detect and repair damages in the DNA as well as to prevent uncontrolled cell division. The molecular events that control the cell cycle are ordered and directional. Each process occurs in a sequential fashion and it is impossible to "reverse" the cycle. The progress of the cell through the cycle is controlled by checkpoints. They verify whether the processes at each phase of the cell cycle have been accurately completed before progression into the next phase. There are three major checkpoints in eukaryotic cell cycle:

- the G_1/S checkpoint (restriction checkpoint) regulates whether cells can enter the process of DNA replication and subsequent division;
- the G_2/M checkpoint — the cell has to check a number of factors to ensure the cell is ready for mitosis (e.g. if all DNA is correctly replicated);
- spindle checkpoint (spindle assembly checkpoint or mitotic checkpoint) — prevents progression of mitosis until all chromosomes are properly attached to the spindle.

7.1.1   CDK and Cyclins

Progression of the cell cycle is catalyzed by cyclin-dependent kinases (CDKs). As the name implies, CDKs are activated by a special class of proteins called **cyclins**. CDKs level remain relatively constant, while concentration of cyclins vary in a cyclical fashion during the cell cycle. Without cyclin, CDK has little kinase activity; only the cyclin-CDK complex is an active kinase. Different complexes are active in different phases of the cell cycle (Fig. 7.2). CDKs phosphorylate their substrates on serines and threonines (see Fig. 2.10c, ch. 2.4.1) triggering specific events during cycle division such as microtubule formation and chromatin remodeling. CDKs and cyclins are large group of proteins. They are present in all known eukaryotes. In mammals different CDKs are denoted by Cdk with consecutive number, and cyclins — with consecutive (capital) letters of the alphabet.

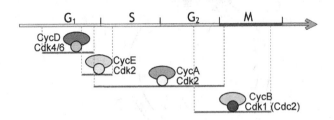

Fig. 7.2 Cyclin-CDK complexes involved in different phases of the cell cycle in mammals. The red line shows the stages of the cycle, in which the individual complexes are present in the cell. Cell cycle begins after the creation of complex of cyclin D (CycD) with Cdk4 or Cdk6. The final stage is catalyzed by a complex of cyclin B (CycB) with Cdk1 (named also Cdc2). This complex is known as **MPF** (maturation promoting factor or mitosis promoting factor).

7.1.2   Mechanism of Initiation and Termination of the S Phase

In the S phase DNA replication takes place. It is triggered when all needed proteins appear in the cell (ch. 4.1). In mammals, the expression of these proteins is activated by the **E2F** transcription factor, which in turn is regulated by the pRB (retinoblastoma) protein (Fig. 7.3). Binding of pRB to E2F inactivates its transcriptional activity, and thus inhibits DNA replication. In late G_1 phase Cyclin D-Cdk4 complex phosphorylates pRB at the specific sites, leading to release of E2F. E2F activates genes needed to start DNA synthesis. It also stimulates the production of cyclin E, cyclin A and its own. Complex Cyclin E-Cdk2 can also phosphorylate pRB accelerating DNA synthesis phase. On the other hand, the complex of Cyclin A-Cdk2 can phosphorylate E2F, inhibiting its transcriptional activity, resulting in termination of DNA replication. It should be noted that the complex Cyclin E-Cdk2 cannot phosphorylate E2F. If that would happen, DNA replication could end prematurely.

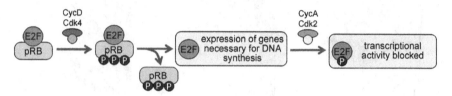

Fig. 7.3 The initiation and termination of S phase

The pRB protein plays a key role in the regulation of cell cycle — it is called "master brake of cell division". It is a tumor suppressor protein that is dysfunctional in many types of cancer.

7.2   Mitosis

Mitosis is the eukaryotic cell division that in conjunction with cytokinesis (cytoplasm division) leads to the formation of two genetically equivalent daughter cells (Fig. 7.4, Fig 1.8). There are many cells (notably among the fungi and slime moulds) where mitosis and cytokinesis occur separately, forming single cells with multiple nuclei.

The period between mitoses is called the **interphase** (includes G_1, S and G_2 phases). In the G_2 phase DNA is already duplicated. The number of chromosomes is 4n. During interphase, chromosomes are not visible in light microscopy, because the chromatin is dispersed in the nucleus.

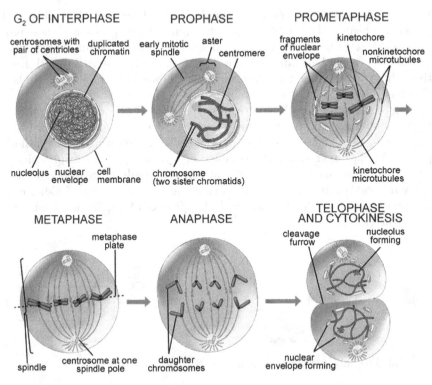

Fig. 7.4 Stages of mitosis on an example of animal cell (to simplify only two pairs of homologous chromosomes are shown)

1. Prophase: chromatin condenses, organizing in chromosomes. Two centrioles migrate to the opposite poles of the cell. From each centriole microtubules spread, forming a structure called the mitotic spindle. At the end of prophase nuclear membrane disintegrates into small vesicles.
2. Metaphase: chromosomes align in the middle of the cell between two centrioles. They are guided by the microtubules attached to the kinetochore present in the centromere of each chromosome.
3. Anaphase: Sister chromatids separate and move away from each other. Separation takes place simultaneously in all chromosomes. Crucial role in this process is played by microtubules associated with the kinetochore.
4. Telophase: Chromosomes begin to unwind — they become less condensed. Spindle disappears and nuclear membrane appears. Cell extends and finally divides into two daughter cells (this process is called cytokinesis).

Altered from:
http://www.upt.pitt.edu/ntress/Bio1_Lab_Manual_New/mitosis_and_meiosis.htm

7.2.1 Initiation of Mitosis

Morphological changes in the early stages of mitosis are caused mainly by the phosphorylation of group of related proteins catalyzed by mitosis promoting factor (MPF — a complex of cyclin B with Cdk1). Activated MPF can phosphorylate histones H1 leading to chromatin condensation. MPF phosphorylates also nuclear lamins (filament proteins of nuclear lamina), which breaks the nuclear membrane. The structure of microtubules is dependent on microtubule-associated proteins (MAP), which can also be phosphorylated by MPF, what leads to the formation of the mitotic spindle. MPF activity is controlled by phosphorylation (Fig. 7.5).

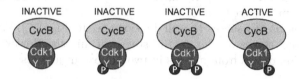

Fig. 7.5 Possibilities of MPF (CycB-Cdk1 complex) phosphorylation and its activity. Phosphorylation can occur on tyrosine (Y) or threonine (T). MPF is active only when it is phosphorylated at specific threonine.

7.2.2 Mechanism of Mitosis Termination

For the mitotic exit, a degradation of the mitotic cyclins is required. This is triggered by the anaphase promoting complex (APC), a multi-subunit complex which contains ubiquitin ligase activity. An inactive APC stabilizes cyclin A and cyclin B securing completion of DNA synthesis and progression through G_2 phase. When mitotic spindle is properly assembled, the APC is activated by phosphorylation catalyzed by the MPF (Fig. 7.6). Mitotic cyclins

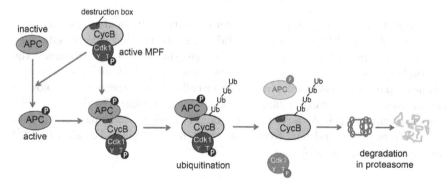

Fig. 7.6 Degradation of cyclin B leading to inactivation of MPF in negative feedback loop

contain a special sequence called destruction box. It is recognized by phosphorylated APC that catalyzes attachment of ubiquitin molecules. Resulting cyclins proteolysis causes the inactivation of cyclin-dependent kinases (e.g. Cdk1, which is the part of MPF). In the absence of active MPF, lamins and other proteins are dephosphorylated by phosphatases. This leads to restoration of nuclear membrane and structures characteristic for interphase cell.

7.3   Meiosis

Meiosis is a special type of division that leads to formation of four genetically nonequivalent daughter cells containing half the number of chromosomes of the parental cell (that is 1n). By meiosis the germ cells, i.e. female egg cells and male sperm, are created.

During meiosis two successive cell divisions occur (Fig. 7.7). Each division consists of four stages similar to the stages of mitosis. However, the first meiotic division differs from mitosis in two important aspects:

1. In the metaphase I, each pair of sister chromatids aligns together with its homologous pair, forming two lines of sister chromatids in the equatorial plane (in mitosis only one line is formed). This alignment is called synapsis. At this stage **crossing-over** occurs, i.e. exchange of fragments of sister chromatids between homologous chromosomes (ch. 7.4.1).
2. In the anaphase I, each pair of sister chromatids moves away from its homologous pair, but the sister chromatids are not separated.

After the first division the number of chromosomes in each daughter cell is 2n. The second meiotic division is similar to mitosis, except that there is no replication of DNA before the division. As a result of divisions haploid cells are obtained (1n).

7.3.1   Independent Chromosome Segregation

In the metaphase I of meiosis, homologous pairs of duplicated chromosomes (each consisting of two sister chromatids) line up together side by side. One chromosome in each pair comes from the egg (the maternal homologue), the second one comes from the sperm (paternal homologue). Homologous chromosomes separate into two different daughter cells. Because pairs of homologous chromosomes are arranged randomly they are distributed to daughter cells in a random manner. As a result, each daughter cell contains part of chromosomes from father and part from mother. In human cells 8.4 million $(= 2^{23})$ combinations are possible.

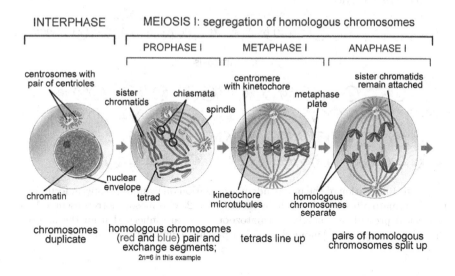

INTERPHASE   MEIOSIS I: segregation of homologous chromosomes

PROPHASE I   METAPHASE I   ANAPHASE I

centrosomes with
pair of centrioles

sister
chromatids   chiasmata

centromere
with kinetochore   metaphase
plate

sister chromatids
remain attached

spindle

nuclear
envelope

chromatin   tetrad

kinetochore
microtubules

homologous
chromosomes
separate

chromosomes
duplicate

homologous chromosomes
(red and blue) pair and
exchange segments;
2n=6 in this example

tetrads line up

pairs of homologous
chromosomes split up

MEIOSIS II: segregation of sister chromatids

TELOPHASE I
and CYTOKINESIS   PROPHASE II   METAPHASE II   ANAPHASE II   TELOPHASE II
and CYTOKINESIS

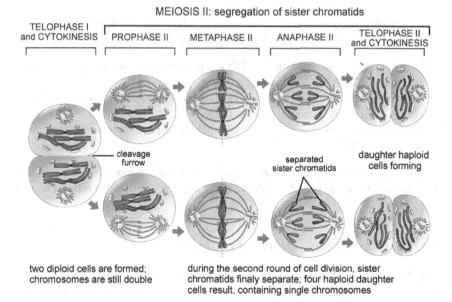

cleavage
furrow

separated
sister chromatids

daughter haploid
cells forming

two diploid cells are formed;
chromosomes are still double

during the second round of cell division, sister
chromatids finaly separate; four haploid daughter
cells result, containing single chromosomes

Fig. 7.7 Scheme of meiosis (to simplify only three pairs of homologous chromosomes are shown)
Altered from:
http://www.upt.pitt.edu/ntress/Bio1_Lab_Manual_New/mitosis_and_meiosis.htm

primary spermatocytes

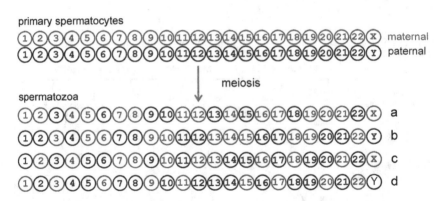

Fig. 7.8 Illustration of the independent segregation on example of sperm formation from spermatocytes, which divide by meiosis. For 23 chromosomes (represented here as circles) present in a single spermatozoon some are inherited from the mother, the other from father. Among 8.4 million possible combinations four are shown.

7.4 DNA Recombination

DNA recombination refers to the process in which a DNA fragment is transferred from one DNA molecule to another. The most common are:

1. Homologous recombination (or DNA crossover) — occurs between two homologous DNA molecules.
2. Site-specific recombination — occurs between two specific, identical DNA sequences present in nonhomologous DNA.
3. Transpositional recombination — when the mobile elements are integrated into target DNA.

Homologous recombination often occurs during meiosis or is used for DNA repair. Other types of recombination are not specifically related to cell division.

7.4.1 Homologous Recombination

Homologous recombination occurs between two homologous, that is identical or similar DNA molecules. It occurs during mitosis (where it is used to repair the double-stranded breaks in DNA) and prophase I of meiosis (pachytene), where it leads to the formation of new combinations of DNA sequences (crossing-over). Crossing-over is a process of great importance for genetic diversity, because it leads to the formation of daughter cells with genotype different than the parental cells. It consists in formation of connections

— chiasmata — between chromatids of neighboring homologous chromosomes, and then separating those connections, so that the exchange of their
segments occurs (Fig. 7.9).

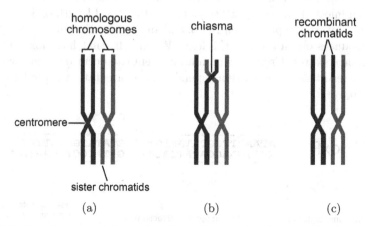

Fig. 7.9 Crossing-over: (a) a pair of homologous chromosomes (each containing
two sister chromatids) aligned side by side; (b) two homologous arms join together
at one point to form chiasma; (c) arms exchange DNA fragment from chiasma to
the end of chromosome

Homologous recombination is used in molecular biology as a technique for
introducing genetic changes into the organism. Thanks to the ability to perform homologous recombination in mammalian cells, the ability to maintain
the culture of embryonic stem cells (ES) or and obtaining chimeric animals,
in 1989 the first organism with an inactivated gene (gene knock-out) was
created. These works have been awarded with the Nobel Prize in Physiology
or Medicine in 2007 (to Oliver Smithies, Mario Capecchi and Martin Evans).

7.4.2 Site-Specific Recombination

Site-specific recombination occurs between two DNA molecules within a
short, specific DNA sequences showing only a limited degree of homology.
For the first time such recombination was observed during the integration
of phage λ DNA into the genome of *E. coli*. Both DNA molecules contain
sequence 5' — TTTATAC — 3' (called a binding site) allowing joining of two
strands of DNA according to the rules of base pairing. Next, integrase enzyme
catalyzes the formation of two single breaks (one on each DNA molecule).
After translocation of free ends, integrase cuts the other two strands and the
phage DNA is integrated into the genome.

Site-specific recombination is very specific, fast, efficient and takes place
without cofactors. It is used as a tool for genetic engineering: it enables to

manipulate with genetic material for testing of gene functions. In common use is so called Cre-loxP system. The Cre (*causes recombination*) recombinase from P1 phage recognizes a sequences called loxP (*locus of crossing-over P1*). Its natural function is the separation of joined together phage genomes created in infected *E. coli*. LoxP sequence, with a total length of 34 bp, consists of two inverted repeats each 13 bp, and an asymmetric 8 bp core sequence that introduces orientation to the loxP. When cells that have loxP sites in their genome express Cre, a recombination event can occur between two loxP sites. The result of recombination depends on the orientation of the loxP sites (Fig. 7.10).

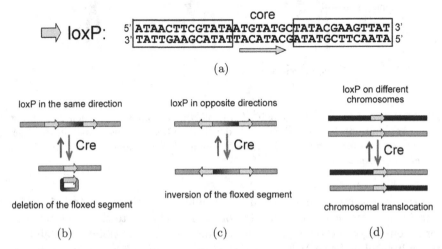

(a)

(b)   (c)   (d)

Fig. 7.10 Characteristics of Cre-loxP system: (a) the loxP structure; (b) Cre recombinase cuts out the DNA fragment floxed by two loxP sequences if they are in the same orientation in the linear DNA molecule; (c) when loxP are placed in opposite directions, the Cre recombinase leads to inversion of the floxed DNA fragment; (d) when loxP are located on two different linear DNA molecules, the Cre recombinase exchanges distal sequences. Recombination reaction can occur in both directions (red arrow). LoxP sequences — marked with a yellow block arrow.

7.4.3   Transposition

Transposition, i.e. relocation of mobile elements from one place in the genome to another, can proceed in two ways: by direct transfer of DNA or through RNA (ch. 5.9). Mobile DNA elements are called transposons, while those transferred through RNA — retrotransposons (Fig. 7.11 and 7.12). Mobile elements are essentially molecular parasites, which appear to have no specific function in the biology of their host organisms, but exist only to maintain themselves. They constitute a significant portion of the genomes of many eukaryotes.

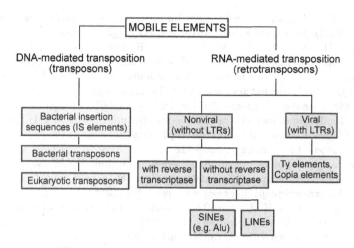

Fig. 7.11 Classification of mobile elements

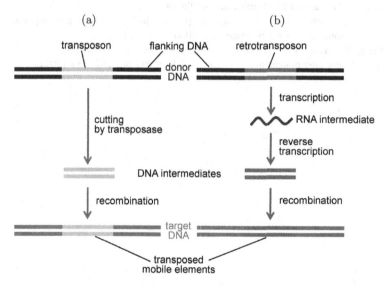

Fig. 7.12 The mechanism of transposition: (a) transposon is cut out from the initial DNA by a special enzyme (in bacteria it is a transposase with nuclease and ligase activity), and then is inserted into the target DNA; (b) retrotransposons are first transcribed from the initial DNA and the resulting RNA is "rewritten" to DNA by reverse transcriptase and then inserted into the target DNA

References

1. King, M.W.: Eukaryotic Cell Cycles. In: themedicalbiochemistrypage.org, LLC (1996-2012), `http://themedicalbiochemistrypage.org/cell-cycle.php`
2. Lodish, H., Berk, A., Zipursky, S.L., et al.: Regulation of the Eukaryotic Cell Cycle. In: Molecular Cell Biology, 4th edn. W.H. Freeman, New York (2000)
3. Pray, L.: Functions and utility of Alu jumping genes. Nature Education 1(1) (2008), `http://www.nature.com/scitable/topicpage/functions-and-utility-of-alu-jumping-genes-561`
4. Pray, L.: Transposons: The jumping genes. Nature Education 1(1) (2008), `http://www.nature.com/scitable/topicpage/transposons-the-jumping-genes-518`
5. Pray, L.: Transposons, or jumping genes: Not junk DNA? Nature Education 1(1) (2008), `http://www.nature.com/scitable/topicpage/transposons-or-jumping-genes-not-junk-dna-1211`
6. Pray, L., Zhaurova, K.: Barbara McClintock and the discovery of jumping genes (transposons). Nature Education 1(1) (2008), `http://www.nature.com/scitable/topicpage/barbara-mcclintock-and-the-discovery-of-jumping-34083`
7. Sullivan, M., Morgan, D.O.: Finishing mitosis, one step at a time. Nat. Rev. Mol. Cell Biol. 8, 894–903 (2007)
8. Tang, Z., Hickey, I. (eds.): Cell Cycle and Cell Division. In: Miko, I. (ed.) Cell Biology. Nature Education (2011), `http://www.nature.com/scitable/topic/cell-cycle-and-cell-division-14122649`
9. Tsai, J.H., McKee, B.D.: Homologous pairing and the role of pairing centers in meiosis. Journal of Cell Science 124, 1955–1963 (2011)
10. Wang, Y., Yau, Y.Y., Perkins-Balding, D., et al.: Recombinase technology: applications and possibilities. Plant Cell Reports 30, 267–285 (2011)

Chapter 8
Cell Signaling Pathways

Cells have an ability to perceive and respond to their microenvironment. External stimuli activate internal signaling pathways that regulate the cell activity. Most signaling pathways function to transfer information from the cell surface to effector systems.

External signals can be transmitted into the cell by several ways (Fig. 8.1):

1. As a hydrophobic molecules directly through the cell membrane;
2. Through ion channels linked receptors;
3. Through receptors associated with G protein;
4. Through receptors which are enzymes or are associated with enzymes;
5. Others (e.g. non-catalytic receptors).

In many cases, an external signal reaches the nucleus and affects gene expression.

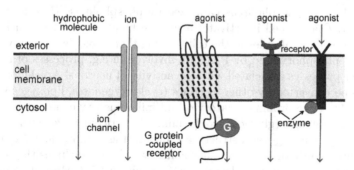

Fig. 8.1 Main pathways of signal transduction into the cell

8.1 Signaling through Hydrophobic Molecules

Hydrophobic molecules, such as nitric oxide (NO), arachidonic acid, or steroids, which play an important role in intracellular signaling, can move into or from the cell directly through the cell membrane. In contrast to other signaling cascades within a single cell, NO or arachidonic acid, formed under the influence of external signals, can leave the cell and influence the neighboring cells.

Nitric oxide is formed from L-arginine with the participation of nitric oxide synthase (NOS). In target cells, NO stimulates cytoplasmic guanylate cyclase to produce cyclic GMP (cGMP). Cyclic GMP (Fig. 8.2) is a secondary messenger which regulates the function of many enzymes and ion channels, e.g. it induces smooth muscle relaxation. cGMP is quite rapidly converted to GMP by phosphodiesterase (PDE). Phosphodiesterase inhibitor is a component of a known drug used to treat erectile dysfunction (Viagra).

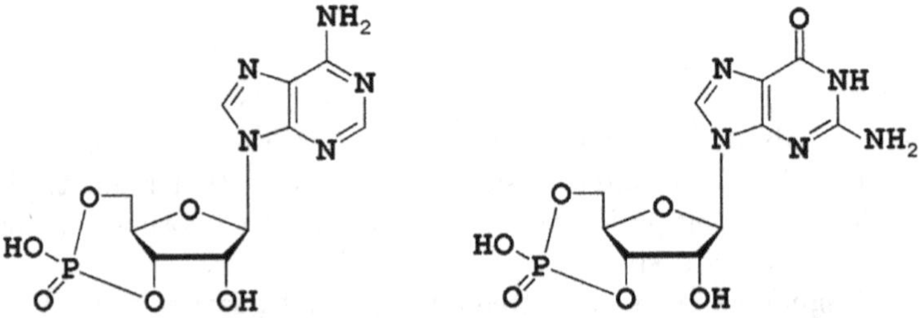

cAMP   cGMP

Fig. 8.2 Structure of cyclic nucleotides: cAMP and cGMP, which are among the most important secondary messengers in signaling cascades

Arachidonic acid (AA or ARA) is a polyunsaturated fatty acid that is present in the membrane phospholipids of the body's cells. In addition, it is involved in cellular signaling as a lipid second messenger. It is formed by hydrolysis of phospholipids catalyzed by phospholipase (Fig. 8.3). After reaching the target cell it activates a protein kinase C (PKC), which phosphorylates some molecules in the cell altering their biological activity. Many molecules phosphorylated by PKC are involved in e.g. processes of learning and other processes associated with the activity of neurons. Arachidonic acid can also be converted by other enzymes (cyclooxygenases, lipoxygenases) to biologically active proinflammatory or antiinflammatory compounds.

Steroid hormones (glucocorticoids, mineralocorticoids, androgens, and female sex hormones; general structure shown in Fig. 1.6, in ch. 1.2.2) belong to the superfamily of fat soluble hormones, which also includes thyroid hormones, retinoids (vitamin A derivatives), and vitamin D3. Most of them enter the cell where they bind to their receptors present in the cytoplasm or the nucleus (only some steroid hormones bind to specific protein receptors on the cell membrane). Steroid hormone receptor usually forms a dimeric form after ligand (hormone in this case) binding. The main role of fat soluble hormones is to regulate transcription: if ligand-receptor complex is formed in the

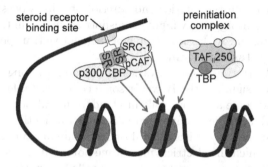

Fig. 8.3 Production of arachidonic acid from phospholipid by phospholipase A_2 (PLA$_2$)

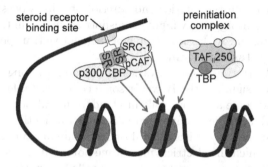

Fig. 8.4 The mechanism of activation of transcription by steroid hormone receptors. Steroid hormone receptor (SR) after joining of ligand (steroid hormone) forms a dimer, which after binding to DNA operates together with steroid receptor coactivators (SRC) and histone acetyltransferases (p300/CBP and p300/CBP-associated factor - pCAF), what stimulates transcription. In the absence of ligand some steroid receptors bind co-repressors (for example SMRT and NCoR) and histone deacetylases, which leads to inhibition of transcription.

cytoplasm, it must be transported to the nucleus, where it acts as a transcription factor. It binds to specific sequences present in regulatory regions of genes and activates or inhibits the transcription (Fig. 8.4).

8.2   Signaling through Ion Channel Linked Receptors

Ion channel linked receptors are ion-channels themselves. It is a large family of multipass transmembrane proteins. They are involved in rapid signaling events. They regulate the flow of ions across the membrane in all cells (ch. 2.4.2.3). Hydrophilic pore (gate) inside the channel is opened or closed depending on external factors. Ion channels may be classified by gating, i.e. what opens and closes the channels. They are divided into three main groups: voltage-gated ion channels, ligand-gated ion channels, and mechanosensitive ion channels. Due to the selectivity, that is ability to pass specific types of ions, channels are divided into cationic and anionic. When the channels are even more "specialized" - are defined as sodium, potassium, etc. It should be noted that the term e.g. "sodium channel" means only that sodium ions are passed preferentialy. However, other cations could also pass through this channel.

All cells have a resting potential: an electrical charge across the plasma membrane, with the interior of the cell negative with respect to the exterior. Sodium, potassium and chloride ions are the most important for maintaining of the resting potential. Usually the concentration of sodium and chloride ions outside the cell is higher than inside the cell, while the concentration of potassium ions is higher inside the cell. Free diffusion of ions occurs through the cell membrane. Maintaining a constant difference of ions concentration between the interior and exterior of cells is possible thanks to the active (i.e. requiring energy input) transport occurring in the opposite direction than diffusion. A classic example of such active transport mechanism is the **sodium-potassium pump**.

The size of the resting potential varies. It can go for a long period of time without changing significantly. However some cells (excitable cells: neurons, muscle cells, and some secretory cells in glands), in addition to maintaining resting potential, are able to change rapidly and transiently their membrane potential in response to environmental or intracellular stimuli. Such short-lasting event in which the electrical membrane potential of a cell rapidly rises and falls, following a consistent trajectory, is called an **action potential**. It is formed in excitable cell when its membrane potential exceeds a certain limit value. Repeated generation of action potentials, that enables the flow of sodium and potassium ions in the direction consistent with the difference in concentrations, would lead to compensation of extracellular and intracellular concentrations of these ions. In all excitable cells there is mechanism of active transport, pumping ions against concentration differences and thus keeping the concentration of ions at a constant level.

Calcium ions also play an important role in the cell. They are involved in signal transduction pathways, where they act as a second messenger, in neurotransmitter release from neurons, contraction of all muscle cell types, and fertilization. Many enzymes require calcium ions as a cofactor. Extracellular calcium is also important for maintaining the potential difference across excitable cell membranes, as well as proper bone formation.

The intracellular calcium level is kept relatively low with respect to the extracellular fluid. Inside the cell, calcium ions are accumulated in the smooth endoplasmic reticulum and mitochondria. Calcium ions can enter into the cytoplasm either from outside the cell through the cell membrane via

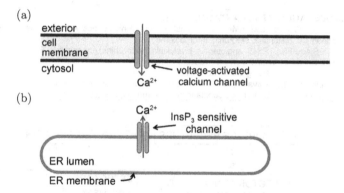

Fig. 8.5 Two types of calcium channels: (a) voltage-gated calcium channel located in the cell membrane; (b) ligand-gated calcium channel located in the membrane of endoplasmic reticulum (ER). InsP3 — triphosphoinositol.

voltage-dependent calcium channels or may be released from the endoplasmic reticulum by a ligand-dependent channels (Fig. 8.5).

8.3 Signal Transduction via Receptors Associated with G Protein

Many signaling molecules bind to specific receptors on the cell membrane (**the molecule does not pass through the membrane**). Such extracellular receptors span the plasma membrane of the cell, with one part of the receptor on the outside of the cell and the other on the inside. A ligand binding to the outside part stimulates a transmition of the signal into the cell, eliciting a physiological response. The signal can be amplified. Thus, one signalling molecule can cause many responses.

External signal could be transmitted via G protein-coupled receptors (GPCRs). Although GPCRs are classically thought of working only with G protein, they may signal through G protein-independent mechanisms (Fig. 8.6). And also, heterotrimeric G proteins may play functional roles independent of GPCRs.

8.3.1 G Protein-Coupled Receptors

G protein-coupled receptors (GPCRs) are large family of transmembrane proteins (in human – about 950 proteins) acting as receptors for extracellular molecules that activate intracellular signaling cascade. GPCRs mediate a wide variety of biological processes, ranging from neurotransmission and

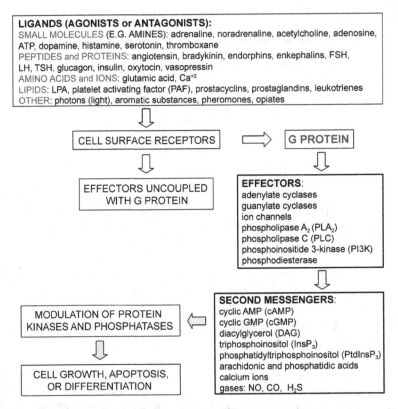

Fig. 8.6 Scheme of signal transduction in cell through the receptors associated with G protein. Activated G protein transfers the signal to one of the effectors what leads to production of the appropriate second messengers.

hormonal control of virtually all physiological responses, to perception of taste, smell, light, and pain. Such receptors are found only in eukaryotes and can bind with a number of ligands listed in the diagram shown in Fig. 8.6 and in ch. 2.4.2.2 (each ligand recognizes its own receptor). Substance (ligand) that binds to the receptor and triggers a response by that cell is called an **agonist**. Whereas an agonist causes an action, an **antagonist** binds to the receptor and blocks it preventing the activation by agonist.

8.3.2 G Proteins

Binding of a ligand to the GPCR causes a conformational change in the receptor. This leads to activation of G protein associated with the receptor, by exchanging its bound GDP for a GTP (Fig. 8.7). G proteins (guanine nucleotide-binding proteins) associated with the GPCRs belong to the family

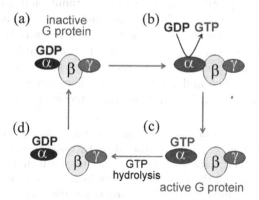

Fig. 8.7 Changes of G protein between inactive and active form: (a) in the inactive form, α subunit is associated with GDP (guanosine diphosphate); (b) interactions between the receptor activated by the agonist and the α subunit lead to the release of GDP. GTP binds in the empty site (its concentration in the cell is much higher than the concentration of GDP); (c) α subunit associated with GTP has small affinity to the $\beta\gamma$ complex, which leads to separation of subunits, so they can activate further signal molecules; (d) α subunit possess GTP-ase activity: it finally hydrolyzes GTP to GDP, what leads to inactivation of the G protein.

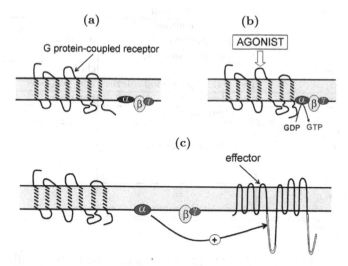

Fig. 8.8 Transmission of signal from the receptor associated with G protein to effector: (a) before the agonist binds to the receptor, three subunits of the G protein form a complex; (b) after binding of agonist, the receptor affecs G protein, changing the GDP in the α subunit for the GTP, so the subunit becomes active; (c) α subunit and $\gamma\beta$ complex are separated and can interact with effector proteins (this figure shows the interaction with adenylyl cyclase responsible for cAMP production). Both, α subunit and $\gamma\beta$ complex, can interact with different effectors (listed in Fig. 8.6).

of heterotrimeric G proteins (composed of α subunit, and β and γ subunits forming the stable complex). They are associated with the inside surface of the cell membrane and serve as a "dashboard" on the way between receptor and effector (Fig. 8.6 and 8.8). G proteins are highly differentiated: over a dozen genes encoding α subunit have been cloned so far (Table 8.1). Also β and γ subunits may exist in different variants. They behave differently in the recognition of the effector, but share a similar mechanism of activation.

Table 8.1 Mammalian G protein α subunits and their effectors. Intracellular communication via G proteins could be affected by bacterial toxins. After binding of the cholera toxin, GTP associated with Gα can not be hydrolyzed to GDP and the G protein remains all the time in the active form. By contrast, pertussis toxin prevents GDP release from the α subunit and G protein is blocked in the inactive state.

Class	Gene variant	Effectors	Sensitivity to toxins
Gα_s	$\alpha_{s(s)}$ $\alpha_{s(L)}$	(+) adenylate cyclase (+) calcium channel (-) sodium channel	cholera
	α_{olf}	(+) adenylate cyclase	cholera
G$\alpha_{i/o}$	α_{i1} α_{i2} α_{i3}	(-) adenylate cyclase (-) calcium channel (+) sodium channel	pertussis
	α_{oa} α_{ob}	(-) calcium channel (+) phospholipase C (+) phospholipase A$_2$	pertussis
	α_{t-r} α_{t-c}	(+) cGMP phosphodiesterase	cholera and pertussis
	α_g	(+) phospholipase C	pertussis
	α_z	(-) adenylate cyclase	
G$\alpha_{q/11}$	α_q α_{11} α_{14} $\alpha_{15/16}$	(+) phospholipase C	
G$\alpha_{12/13}$	α_{12} α_{13}	(+) small G protein	

Notes: (+) activation; (-) inhibition

G proteins were discovered by Alfred G. Gilman and Martin Rodbell when they investigated stimulation of cells by adrenaline. For this discovery, they won the 1994 Nobel Prize in Physiology or Medicine.

8.3.3 Effectors and Secondary Messengers

Both activated α subunit and released $\beta\gamma$ subunits of the G protein can activate different signaling cascades and effector proteins (listed in the diagram in Fig. 8.6). The main effectors on which α subunit acts are listed in table 8.1 ($\gamma\beta$ complex also can act on the same effectors). Adenylate cyclase, a membrane protein, catalyzes the conversion of ATP to cyclic AMP, one of the most important secondary messengers (Fig. 8.2; ch. 8.5.1). Whereas guanylate cyclases (membrane-bound or soluble forms) catalyze the formation of cGMP. Phospholipase A2 (PLA_2) cleaves phospholipids generating the formation of arachidonic acid (Fig. 8.3). Phospholipase C (PLC) also cleaves phospholipids, but in a different position than PLA_2, leading to the formation of diacylglycerol (DAG) (Fig. 8.9). Among the variety of PLC (β, γ, δ, and others), PLC-β is activated by G protein.

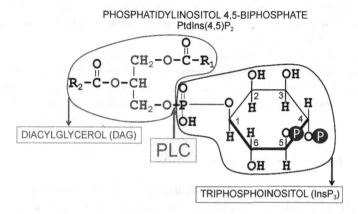

Fig. 8.9 An example of the reaction catalyzed by phospholipase C (PLC): hydrolysis of phosphatidylinositol diphosphate generates the formation of secondary transmitters: diacylglycerol (DAG) and triphosphoinositol ($InsP_3$, IP_3). DAG remains in the membrane and activates protein kinase C (PKC)

Second messengers produced by effectors relay signals to target molecules inside the cell, in the cytoplasm or nucleus. They greatly amplify the strength of the signal.

There are three basic types of secondary messenger molecules:

– hydrophilic (i.e. water-soluble), like cAMP, cGMP, $InsP_3$, and calcium ions, that are located within the cytosol;
– hydrophobic (i.e. water-insoluble), like diacylglycerol, and phosphatidylinositols, which are membrane-associated and diffuse from the plasma membrane into the intermembrane space where they can reach and regulate membrane-associated effector proteins;
– gases: nitric oxide (NO), carbon monoxide (CO) and hydrogen sulphide (H_2S) which can diffuse both through cytosol, and across cellular membranes.

For discovery of second messengers, Earl Wilbur Sutherland, Jr. won the 1971 Nobel Prize in Physiology or Medicine.

8.4   Signal Transduction through Enzyme Linked Receptors

Enzyme-linked receptors are either enzymes themselves, or are directly associated with the enzymes that they activate. These are usually single-pass transmembrane receptors, with the enzymatic portion of the receptor being intracellular. The majority of enzyme-linked receptors are protein kinases, or associate with protein kinases, mainly tyrosine or serine/threonine kinases. These kinases phosphorylate respectively tyrosine and serine or threonine in target proteins. They are capable of autophosphorylation as well as phosphorylation of other substrates. Protein phosphorylation leads to changes in its enzymatic activity, intracellular localization, or in interactions with other proteins. Protein kinases are also found in the cytosol in the form of enzymes not associated with the receptors.

Abnormal activity of receptor kinases is a common cause of diseases, especially carcinogenesis. Signaling pathways affected by such kinases could serve as targets for cancer therapy. Some drugs that are inhibitors of protein kinases are already approved for treatment, other drugs are in clinical trials stage.

8.4.1   Receptor Tyrosine Kinases

Of the 91 unique tyrosine kinases identified so far, 59 are receptor tyrosine kinases (RTKs). They are transmembrane proteins: the domain with tyrosine kinase activity is located on the cytoplasmic side (together with regulatory domain). Extracellular region contains ligand binding domain. It can be a separate unit connected with the rest of the receptor by disulfide bond. Receptor tyrosine kinases are involved in transmitting signals into the cell, and their activity is extremely important in the regulation of cell division, differentiation, and morphogenesis. The majority of them are receptors for growth factors and hormones like epidermal growth factor (EGF), platelet derived growth factor (PDGF), fibroblast growth factor (FGF), hepatocyte growth factor (HGF), insulin, nerve growth factor (NGF), vascular endothelial growth factor (VEGF), macrophage colony-stimulating factor (M-CSF), etc.

Binding of ligand to the receptor leads to oligomerization (usually dimerization) of monomeric receptor kinase or stabilization of already existing loosely associated dimers. Thereafter *trans*-autophosphorylation of kinases (i.e. phosphorylation of neighboring kinase in dimer) occurs. As a result of phosphorylation, changes in the spatial structure of the receptor are generated, causing the opening of kinase domain and binding of ATP in catalytic center (Fig. 8.10). Activated tyrosine kinase phosphorylates specific target proteins - often they are also enzymes, such as cytoplasmic tyrosine kinases.

This starts the kinases cascade, in which further types of kinases are phosphorylated and activated. As a result, transcription factors coud be activated or inactivated leading to changes in the expression of specific genes.

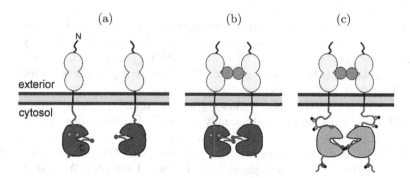

Fig. 8.10 General model of regulation of the activity of receptor tyrosine kinases: (a) in the absence of ligand the tyrosine kinase domain (dark green) of a receptor maintains the basal, low activity due to inhibitory interactions with the region adjacent to the membrane (red) and/or the carboxyterminal tail (violet). In addition, the activation segment (brown) has a structure that blocks an access to the catalytic center; (b) after binding of ligand (orange) and dimerization of the extracellular domains (yellow), the cytoplasmic domains are juxtaposed, which facilitates the trans-autophosphorylation of tyrosine residues (shown as circles) in the region adjacent to the membrane, in the activation segment and in the carboxy-terminal tail; (c) after phosphorylation (black dots) and reconfiguration of the inhibitory segments, the kinase domains become fully active (light green). Then phosphotyrosines interact with proteins containing SH2 (Src Homology 2) domains or phosphotyrosine binding domains. This initiates a series of events which eventually result in altered patterns of gene expression or other cellular responses.
Altered from: http://www.nature.com/nrm/journal/v5/n6/fig_tab/nrm1399_F1.html
Hubbard SR (2004) Juxtamembrane autoinhibition in receptor tyrosine kinases. Nature Reviews Molecular Cell Biology 5, 464-471.

Many ligands of receptor tyrosine kinases are multivalent. Also some tyrosine receptor kinases may form heterodimers with similar but not identical kinases. It is the reason why the response to the extracellular signal can be highly varied.

Cell must have a possibility to stop the signal response. Constant activity of receptors, e.g. growth factors, would lead to the uncontrolled division and consequently to carcinogenesis. Ligand-receptor complex, in case of receptor tyrosine kinases, is absorbed and destroyed through receptor endocytosis.

A class of tyrosine kinases recruited to the receptor just after connection of ligand to it is involved in signaling cascades activated among others by cytokines. One of such kinases is JAK kinase (Janus kinase) regulating activity of STAT proteins (ch. 8.5.3). **Cytokines** are protein molecules that affect growth, proliferation and stimulation of hematopoietic cells and cells involved in immune response.

8.4.2 Receptor Serine/Threonine Kinases

At least 125 of more than 500 human protein kinases are serine-threonine kinases (STKs). Their activity is regulated by cAMP/cGMP, diacylglycerol (DAG), Ca^{2+}/calmodulin, or DNA damage. STKs select specific residues to phosphorylate on the basis of residues that flank the phosphoacceptor site, which together comprise the consensus sequence.

Receptor serine/threonine kinases transmit mainly signals induced by growth factors belonging to the family of TGFβ (transforming growth factor β). They exist as heterodimers consisting of type I and type II receptor. Ligand binding domain is present in type II receptor. Type II receptor binds ligand and then recruits and phosphorylates type I receptor. Activated type I receptor phosphorylates specifically SMAD proteins, which are transported to the nucleus where regulate gene expression.

8.5 Intracellular Signaling Pathways

There are a large number of intracellular signaling pathways responsible for transmitting information within the cell. The majority respond to external stimuli, which are received by receptors embedded in the plasma membrane. These receptors then transfer information across the membrane using a variety of transducers and amplifiers that engage a diverse repertoire of intracellular signaling pathways. The other categories are the pathways that are activated by signals generated from within the cell. All of these signaling pathways generate an internal messenger that is responsible for relaying information to the sensors that then engage the effectors that activate cellular responses. Signaling pathways often interact with each other. Interactions (positive and negative) of second messengers systems (so called cross-talk) decide on the final activity of different pathways and biological response. Signaling pathways are dynamic and can proceed differently in different cell types. Several examples of the better known cell signaling pathways are outlined below.

8.5.1 cAMP-dependent Signaling Pathway

Cyclic AMP is involved in the activation of protein kinases and regulates the effects of adrenaline and glucagon. It also binds to and regulates the function of ion channels dependent on cyclic nucleotides and other factors. Protein kinase A (PKA) is the main target of cAMP. PKA is present in the cytoplasm in an inactive form as tetramer (R2C2), composed of two catalytic subunits (C) and two regulatory subunits (R). Usually, it is located in the perinuclear space (thanks to the AKAP protein - membrane-associated archoring protein). Cyclic AMP binds to regulatory subunits, which leads to the release of PKA catalytic subunits. Active PKA can phosphorylate among others: enzymes associated with the synthesis of neurotransmitters and metabolism of cyclic nucleotides, neurotransmitter receptors, ion channel proteins, proteins involved in regulation of transcription and translation, or cytoskeletal proteins. Processes dependent on cAMP/PKA (as well as cGMP/PKG) are mostly short-term processes, because rapid dephosphorylation of these proteins occurs. A more prolonged cellular responses are associated with the influence of cAMP on gene expression (Fig. 8.11). Released PKA catalytic subunits migrate to the nucleus (by passive diffusion), where they phosphorylate CREB protein (cAMP response element-binding protein). This leads to transcription of genes containing CRE sequence (cAMP response element) in the promoters (ch. 5.3.1).

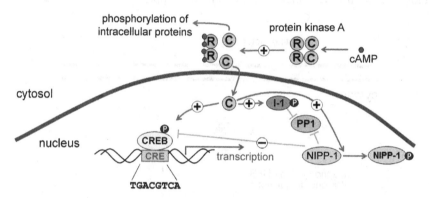

Fig. 8.11 Example of the complexity of processes dependent on cAMP/PKA. cAMP binds to two sites in each of the regulatory subunits (R) of PKA (protein kinase dependent on cAMP). Released catalytic subunits (C) phosphorylate serines and threonines in the target proteins. Long-term PKA activity is associated with effects on the gene expression through CREB. CREB activity can also be regulated by other enzymes: calmodulin, nerve growth factor NGF, and phosphatases. PKA phosphorylates at the same time CREB, I-1 protein and NIPP-1 protein. I-1 in phosphorylated form is an inhibitor of protein phosphatase PP1, while NIPP-1 in dephosphorylated form is an inhibitor of PP1, thus in phosphorylated form has activity opposite to the phosphorylated I-1. PP1 activity is a function of activities of phosphorylated proteins I-1 and NIPP-1. If PP1 is active, it dephosphorylates proteins phosphorylated by PKA.

8.5.2 NF-kappaB Signaling Pathway

NF-κB transcription factor can be activated by many signals, including TNF-α (tumor necrosis factor α), interleukin 1 (IL-1), compounds triggering mitosis (mitogens) in T and B cells, bacterial lipopolysaccharides (LPS), viral proteins, and others. It controls the expression of genes involved in immune response, apoptosis, and cell cycle regulation. Therefore, the abnormal regulation of NF-κB may cause an inflammation, autoimmune diseases, viral infections, and carcinogenesis. In mammals, the family of NF-κB includes five proteins: NF-κB1 (or p50), NF-κB2 (or p52), RelA (or p65), RelB, and c-Rel. All of them possess conserved in evolution Rel domain, responsible for dimerization, binding to DNA, and for binding of IκB (inhibitor of NF-κB). NF-κB works only as a dimer. The most common active form of NF-κB contains subunits p50 or p52, and p65.

In the cytoplasm, NF-κB is inhibited by IκB. Activation signal transmitted from receptor (e.g. after binding of TNF-α to its receptor) leads to phosphorylation of the inhibitor by the IKK kinase (IκB kinase). This results in ubiquitination and degradation of IκB in the proteasome. Released NF-κB migrates to the nucleus, where it activates transcription of various genes, including its inhibitor, which within one hour reappears in the cell (Fig. 8.12).

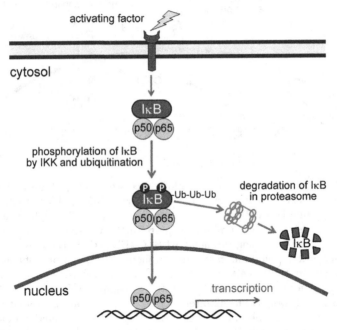

Fig. 8.12 General model of NF-κB activation

8.5.3 JAK-STAT Signaling Pathway

The JAK-STAT system consists of three main components: a receptor, JAK, and STAT. The receptors do not possess catalytic kinase activity, thus they rely on the JAK family of tyrosine kinases. JAK is short for Janus Kinase. STAT proteins (Signal Transducer and Activator of Transcription) (family of seven proteins) are proteins both transmitting signal and activating transcription. The inactive form is a monomer and is in constant motion between the cytoplasm and the nucleus awaiting for activating signal. The JAK-STAT system is a major signaling alternative to the second messenger system.

Signal activating the JAK-STAT pathway comes from the membrane receptors. They can be activated by interferon, interleukin, growth factors, or other chemical messengers. The receptors exist as paired polypeptides and JAKs are associated with intracellular domains. Ligand binding enables a conformational change of the receptor and then JAK kinase autophosphorylates itself. The STAT protein then binds to the receptor. STAT is phosphorylated, forms a dimer (through the SH2 domain) and translocates into the cell nucleus, where it binds to DNA and promotes transcription of genes responsive to STAT (Fig. 8.13). STAT protein is dephosphorylated and inactivated by nuclear phosphatases, and then transported to the cytoplasm.

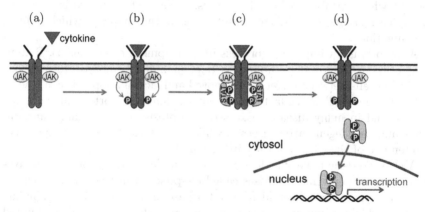

Fig. 8.13 General model of STAT protein activation

8.5.4 MAPK Signaling Pathway

Mitogen-Activated Protein Kinases (MAPKs) are serine/threonine-specific protein kinases. They respond to variety of external signals (mitogens, osmotic stress, heat shock, and proinflammatory cytokines), resulting in either cell growth, differentiation, inflammation, or apoptosis. Generally, the signal is transmitted as follows ("MAPK cascade"):

$$\text{stimulus} \rightarrow \textbf{MAPKKK} \rightarrow \textbf{MAPKK} \rightarrow \textbf{MAPK} \rightarrow \textbf{response},$$

where MAPKK is MAPK kinase, and MAPKKK - MAPKK kinase.

Ligands that activate MAPK's cascade bind receptor tyrosine kinases. Next, signal is transmitted by small monomeric G proteins (sometimes by calcium ions or certain heterotrimeric G proteins) to MAPKKK. MAPK regulates the activities of several transcription factors (e.g. MYC, FOS), what leads to altered transcription of genes that are important for the cell cycle. Another effect of MAPK activation is to alter the translation of mRNA to proteins.

8.6   Apoptosis

Apoptosis, also known as programmed cell death (PCD), is strictly regulated, active process of self-destruction of an individual cell that may occur in multicellular organisms. It is an important cellular process that starts by specific signaling pathways in response to some external and internal stimuli. Morphologically, apoptosis is characterized by chromatin condensation and shrinkage of cell, and then by fragmentation of the nucleus and cytoplasm. As a result, apoptotic bodies surrounded by the cell membrane are formed, which are absorbed by phagocytes. In contrast to apoptosis, cells that die by necrosis swell and disintegrate, releasing the contents, which induces inflammation.

Apoptosis is an important process in development of organisms. Redundant cells are removed by apoptosis (e.g. in human embryo, during digit formation cells between fingers are removed and lack of apoptosis can lead to webbed fingers called syndactyly). Apoptosis is also important in the aging process and in many diseases. Excessive apoptosis occurs in many autoimmune and autodegenerative diseases, while not sufficiently efficient apoptosis is often one of the causes of the carcinogenesis.

A characteristic molecular feature of cells undergoing apoptosis is activation of specific cysteine proteases called **caspases**. Until now more than 10 different caspases have been identified. Caspases are involved in the signaling cascade, in which consecutive enzymes are activated as a result of limited proteolysis. Some of them, called initiator (apical, e.g. caspase-2, -8, -9, and -10), are activated by specific stimulus and start a cascade. Others, called effector (executioner, e.g. caspase-3, -6, and -7), are activated by initiator caspases and catalyze the proteolysis of other proteins essential for cellular functions. Signaling pathway associated with caspases includes:

1. Activation of apical caspases by specific initiatory signals, e.g. changes in membrane receptors status.
2. Activation of effector caspases by initiatory caspases, which cut inactive caspases (procaspases) at specific sites.
3. Specific proteolysis of important cellular proteins by effector caspases. One of the substrate for effector caspases is inhibitor (DFF45/ICAD protein) of apoptotic nuclease (DFF40/CAD). The cleavage and inactivation of

the inhibitor allows nuclease to enter the nucleus and fragment the DNA, which is an irreversible stage of apoptotic cell death.

Signals inducing apoptosis can reach the cell from the outside. Death receptor (extrinsic) pathway of apoptosis is then triggered (Fig. 8.14). So called death ligands (e.g. FasL/CD95L, TRAIL, APO-3L, or TNF) bind with cell membrane specific death receptors (Fas/CD95, DR3, DR4, DR5, TNFR) possessing the death domain (DD). Binding of ligand to receptor induces its trimerization and attachment of adaptor proteins, also containing death domains (FADD, TRADD, RAIDD, DAXX, and others). As a result, initiatory caspases-8 or -10 are activated.

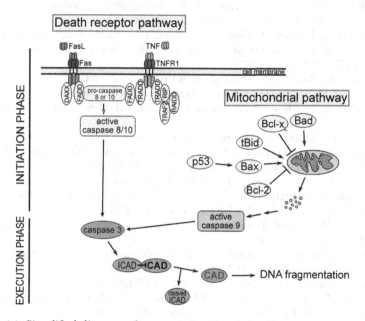

Fig. 8.14 Simplified diagram of transmitting signals leading to apoptosis

Apoptosis can also be caused by intracellular factors, mainly by signals from the mitochondria (mitochondrial or intrinsic pathway of apoptosis). Factors leading to increase of mitochondrial membrane permeability may influence release of proapoptotic factors (such as cytochrome c) into the cytosol. Released cytochrome c, together with APAF1 protein activates the initiator caspase-9, which then activates executioner caspase-3.

In cells undergoing apoptosis the process of cytochrome c (and other apoptotic proteins) release from mitochondria is regulated by a number of supporting proteins from Bcl-2 family. This family consists of three subfamilies. Two of them are formed by proteins present in the outer mitochondrial membrane: pro-apoptotic Bax and Bak proteins, and anti-apoptotic Bcl-2 and

Bcl-xL proteins. These proteins in mitochondrial membrane can form homo-or heterodimers. The third subfamily, including Bid, Bik, and Bim, includes pro-apoptotic proteins present in the cytoplasm. They can interact with membrane proteins from Bcl-2 family and modify their activity. It is believed that membrane proteins from Bcl-2 family can form membrane pores enabling outflow of cytochrome c to the cytoplasm or can interact with proteins forming ion channel VDAC (voltage-dependent anion channel) and modulate its permeability to cytochrome c.

References

1. Balla, T., Szentpetery, Z., Kim, Y.J.: Phosphoinositide signaling: new tools and insights. Physiology (Bethesda) 24, 231–244 (2009)
2. Chen, R.E., Thorner, J.: Function and regulation in MAPK signaling pathways: lessons learned from the yeast Saccharomyces cerevisiae. Biochim. Biophys. Acta 1773, 1311–1340 (2007)
3. Gewies, A.: ApoReview - Introduction to Apoptosis (2003), http://www.celldeath.de/encyclo/aporev/aporev.htm#Index
4. Hayden, M.S., Ghosh, S.: Signaling to NF-kappaB. Genes Dev. 18, 2195–2224 (2004)
5. Hille, B.: G protein-coupled receptor. Scholarpedia 4(12), 8214 (2009), http://www.scholarpedia.org/article/G_protein-coupled_receptor
6. Hille, B.: Ion channels. Scholarpedia 3(10), 6051 (2008), http://www.scholarpedia.org/article/Ion_channels
7. Hubbard, S.R.: Juxtamembrane autoinhibition in receptor tyrosine kinases. Nature Review Molecular Cell Biology 5, 464–471 (2004)
8. King, M.W.: Signal Transduction. In: themedicalbiochemistrypage.org, LLC (1996-2012), http://themedicalbiochemistrypage.org/signal-transduction.php
9. Kroeze, W.K., Sheffler, D.J., Roth, B.L.: G-protein-coupled receptors at a glance. Journal of Cell Science 116, 4867–4869 (2003)
10. Lodish, H., Berk, A., Zipursky, S.L., et al.: Cell-to-Cell Signaling: Hormones and Receptors. In: Molecular Cell Biology, 4th edn. W.H. Freeman, New York (2000)
11. Neitzel, J., Rasband, M. (eds.): Cell Communication. In: Miko, I. (ed.) Cell Biology. Nature Education (2011), http://www.nature.com/scitable/topic/cell-communication-14122659
12. Rawlings, J.S., Rosler, K.M., Harrison, D.A.: The JAK/STAT signaling pathway. Journal of Cell Science 117, 1281–1283 (2004)

Chapter 9
High-Throughput Technologies in Molecular Biology

Over the past few decades, advances in the computer science and informatics have made possible development of new strategies in molecular biology that enable to collect and analyze the data about complex interactions within biological systems. A new inter-disciplinary field of study, called systems biology, has been created. It analyzes and integrates information from genomics (DNA level), transcriptomics (RNA level), proteomics (protein level) and other related fields (e.g. metabolomics and other -omics). Specialized tools to view and analyze biological data and computerized databases to store, organize, and index the data are required. Thus, bioinformatics, the interface between the biological and computational sciences, is indispensable to obtain a clearer insight into the fundamental biology of organisms.

9.1 DNA Sequencing

DNA sequencing is the laboratory method used to determine the exact order of base pairs in a DNA fragment. The DNA sequence contains information necessary for organisms to survive and reproduce, thus reading the sequence is useful in fundamental research into why and how organisms live. With the sequencing of genomes, genomics was developed, which can be defined as the mapping of genes of various organisms by large-scale DNA-sequence analysis.

9.1.1 Chain Termination Sequencing

The first techniques of sequencing were established in the 1970-1980s. One of the major methods is known as chain termination sequencing, dideoxy sequencing, or Sanger sequencing after its inventor Frederick Sanger (Nobel Prize in Chemistry in 1980, shared with Walter Gilbert). The single stranded DNA fragment (which is the sequence to be determined) serves as a template

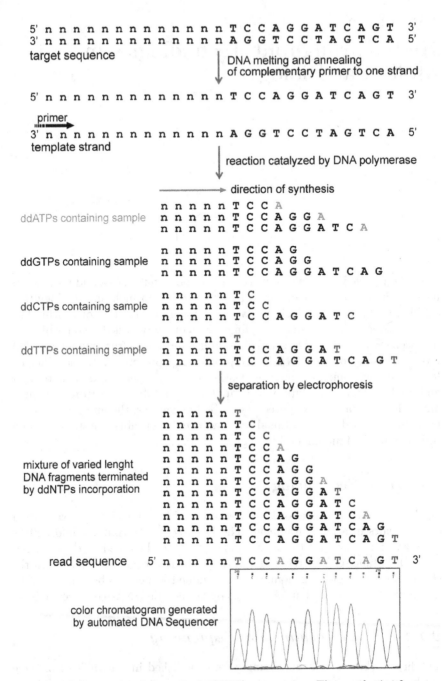

Fig. 9.1 Chain termination method of DNA sequencing. The synthesis of a new chain stops after incorporation of dideoxynucleotide (ddNTP). If ddNTPs are labeled differently, products of the reaction can be analyzed together, if not - samples with each ddNTP should be analyzed separately.

for DNA polymerase, that normally uses deoxynucleotides (dNTPs) to synthesize a new strand starting from the primer complementary to the given sequence (ch. 4.1.1). In the Sanger method, dideoxynucleotides (ddNTPs) - nucleotides missing a hydroxyl (-OH) group at the 3' position (see Fig. 3.2), are added to the reaction. To this position normally a new nucleotide is attached in condensation reaction (Fig. 3.5). If there is no -OH group in the 3' position, the additional nucleotides cannot be added to the chain, thus interrupting chain elongation. In the manual method of sequencing one of dNTPs or primer are labeled (e.g. by radioactive sulfur) and the sample has to be divided into four tubes, each containing all dNTPs and a smaller amount of one of ddNTPs (ddATP, ddTTP, ddCTP or ddGTP). During reaction ddNTPs have a chance to be incorporated in each position in the sequence. Providing that there is enough substrates in the reaction tube, both template and nucleotides, the obtained terminated products will be of all possible lengths, differing in size by one nucleotide. Finally, the mixture of varying length fragments is run on a gel and the sequence can be read from the smallest fragment to the largest (Fig. 9.1).

Since 1986 the process of sequence reading can be done with an automated fluorescence sequencer. Automated sequencing uses fluorescent tags on the ddNTPs (a different dye for each nucleotide). This makes it possible for all four reactions to be run in one lane and increases the speed of the process. A laser constantly scans the bottom of the gel, detecting the bands that move down the gel. The runs are fully automated nowadays, and the gels are replaced by capillaries.

9.1.2 Full Genome Sequencing

Chain termination methods (sometimes referred to as first generation sequencing) can directly sequence only relatively short (300 - 1,000 nucleotides long) DNA fragments in a single reaction. Thus, longer sequences (like genomic DNA) has to be fragmented (with restriction enzymes or mechanical forces) into random pieces. In conventional genome sequencing, DNA fragments are subcloned into vectors (e.g. plasmids) and introduced into bacterial cells to prepare a library covering the whole genome. The transformed cells containing subclones are plated and grown and then the DNA from each one is isolated and sequenced in a capillary sequencer (Fig. 9.2a). The sequence is assembled electronically into one long, contiguous sequence by using overlapping DNA regions. Such method of genome sequencing (called shotgun sequencing technology) was applied to sequence the first human genome. Sequencing of the essentially complete genome (by The Human Genome Project and Celera Genomics) took 13 years and was announced in April 2003. The sequence of the DNA along with sequences of known and hypothetical genes

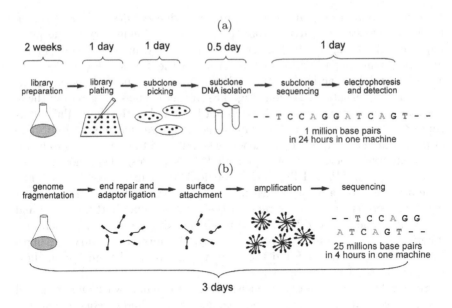

Fig. 9.2 Comparison of conventional chain termination sequencing (a) and massively parallel sequencing (b). Altered from: Mardis E.R. (2006) Anticipating the 1,000 dollar genome. Genome Biol. 7, 112.

and proteins is stored in databases such as GenBank housed by the National Center for Biotechnology Information.

In another method for *in vitro* clonal amplification, genomic DNA fragments first undergo end-repair to provide blunt ends for adaptor ligation. Next, they have specific adaptors ligated to their ends that contain priming sites for PCR (polymerase chain reaction, see ch. 4.1.2) and sequencing. The adaptor-ligated fragments are then hybridized to complementary adaptors that are fixed to a surface (a slide or bead). Then *in situ* PCR amplification is used instead of bacterial amplification. Sequencing reactions of the amplified fragments take place on the surface (Fig. 9.2b). The sequence is visualized base by base simultaneously in all fragments using either luciferase or fluorescence reporting that is detected by a CCD camera. Such technology, named massively parallel genome sequencing (or second generation sequencing), was announced in 2005. It has reduced the cost and increased the throughput of genomic sequencing. Nowadays smaller genomes can be synthesized in one experiment. The technology is widely used also for microRNA screening, gene expression analyzes and ChIP-seq (see ch. 9. 3).

The second generation technologies were initially inferior to classical sequencing in terms of read length (about 35-76 nt for Illumina platform versus about 700 nt for classical sequencing) and single-read error rate (about 1-2% versus less than 0.1%). Rapidly developing third generation technologies (frequently identified with single-molecule sequencing) provide solutions to some of the problems that second generation sequencing faced by simplifying sample preparation, reducing sample mass requirements, and eliminating amplification of DNA templates. There are several different strategies of fast sequencing and a number of public and private companies are competing to commercialize full human genome sequencing. However, we should keep in mind that although full genome sequencing provides raw data on all six billion letters in an individual's DNA, it does not provide an analysis of what that data means or how that data can be utilized in various clinical applications, such as in medicine to help prevent disease.

9.2 Analysis of Gene Expression at RNA Level

The expression of genes can be determined by measuring mRNA levels. For a single gene, Northern blotting can be used. The technique is based on nucleic acids hybridization, that is preferential binding of complementary single-stranded nucleic acid sequences (see ch. 3.2.2). The total RNA is isolated from cells and after separation in a gel (electrophoresis) bound to the membrane. A labeled probe (introduced into the solution in single-stranded form) specific for the examined gene can hybridize and leave the signal only with RNA molecules with a complementary sequence. Nowadays, the most useful method to evaluate the expression of a single gene is the RT-PCR (reverse transcription - polymerase chain reaction) technique (especially real time RT-PCR, which is a quantitative method). Reverse transcriptase first copies RNA to DNA (such DNA created from RNA is called cDNA - complementary DNA), next the examined sequence is amplified in a series of PCR cycles by DNA polymerase using specific primers (ch. 4.1.2; Fig. 4.4).

The complete genome sequencing has opened the way to the investigation of the transcriptional activity of thousands of genes (i.e. the transcriptome) in selected tissues, in one experiment through DNA microarrays. Microarrays are producing massive amounts of data. These data, like genome sequence data, can help us to gain insights into underlying biological processes. The structural analysis of accumulated transcripts in a cell or tissue is the main goal of transcriptomics.

9.2.1 DNA Microarrays

The basic use of microarrays (arrays, chips) is gene expression profiling, thus RNA detection. Microarrays can be used also to detect DNA: in studies of DNA-protein interactions (ch. 9.3), in comparative genomic hybridization, chromatin modifications, SNPs (single nucleotide polymorphisms) detection, and in other applications. Microarrays studies, similarly to Northern blotting, are based on nucleic acids hybridization. The main difference between these two techniques is that in Northern blotting RNA is fixed with a surface and one labeled probe is used in a hybridization solution, while in microarrays multiple probes are attached at fixed locations (spots) to a surface (typically a glass slide) and total RNA is amplified, labeled and introduced into the hybridization solution. There may be tens of thousands of spots on an array, each containing a huge number of identical single stranded DNA molecules of lengths from twenty to hundreds of nucleotides. The spot diameter is of the order of 0.1 mm or smaller. The identity of each spot is known by its position. The whole array occupies no more than a few dozen square centimeters. For gene expression studies, probes from one spot ideally should identify (be complementary to) one gene or one exon in the genome. In practice this is not always so simple and may not even be generally possible due to families of similar genes in a genome.

In microarrays, the probes are oligonucleotides, cDNA or small fragments of PCR products that correspond e.g. to mRNAs. The probes are synthesized prior to deposition on the array surface and are then "spotted" onto the surface. Arrays can also be produced by printing desired sequences by synthesizing directly onto the array surface instead of depositing intact sequences. Printed sequences may be longer (60-mer probes such as the Agilent design) or shorter (25-mer probes produced by Affymetrix). Although oligonucleotide probes are often used in "spotted" microarrays, the term "oligonucleotide array" most often refers to "printed" array.

In microarrays dedicated to expression studies, probes are complementary to sequences of known or predicted genes, to their transcribed regions (preferentially to 3'UTRs, as a regions more divergent even in families of highly

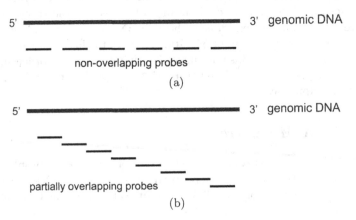

Fig. 9.3 Types of genomic tiling arrays: covering almost the whole (a) or the whole (b) genomic sequence

similar genes). There are also commercially available arrays dedicated to microRNA. Furthermore, DNA microarrays can densely cover a whole genome or a genomic region of interest. They are called genomic tiling arrays (Fig. 9.3). In such arrays usually oligonucleotides are used. To cover a mammalian genome several arrays are necessary (e.g. to interrogate the mouse genome, Affymetrix offers the set of seven arrays containing approximately 45 million oligonucleotide probes). Tiling arrays dedicated to studies of transcription factors interactions with DNA (by ChIP-chip; ch. 9.3) are a single arrays. For example Affymetrix Promoter Array is comprised of over 4.6 million probes (25 bp long) tiled to interrogate over 25,000 known or predicted promoter regions. Each promoter region covers approximately 6 kb upstream through 2.5 kb downstream of 5' transcription start sites.

Oligonucleotide microarrays can be used to detect point mutations (the missing, adding, or changing of a single base) in a known DNA sequence. Single base mismatches do have much more influence on binding to an oligonucleotide sequence compared to much longer cDNA. For example, a small genome can be loaded on a chip as a set of thousands of 20-25 bp long fragments. When a single base pair match exists, the fluorescence intensity decreases significant. This technique gives possibilities to find most of the point mutations (or SNPs) in a known DNA sequence.

9.2.1.1   Two-Color Microarrays

One of the most popular microarray applications allows the comparison of two different samples (e.g. RNA from the same cell type in a "healthy" and "diseased" state) hybridizing them on one array (Fig. 9.4). The total mRNA from the cells in two different conditions is extracted and reverse transcription PCR (RT-PCR) is used to convert the transcripts into cDNA. The cDNAs are usually composed of 500 -2,000 base pairs long. They have to be amplified and labeled with two different fluorophores (commonly used are cyanine 3- or cyanine 5-labeled CTP: abbreviated to Cy3 and Cy5, respectively). Nucleotides labeled with a red fluorescent dye Cy5 can be incorporated into the experimental cDNA (e.g. from cells in a "diseased" state), while nucleotides labeled with a green fluorescent dye Cy3 - into the control cDNA (e.g. from cells in a "healthy" state). Both samples are mixed and after fragmentation allowed to hybridize overnight to the array. Excess hybridization buffer is washed off and the slides are then scanned. Labeled molecules (in this case - products of gene expression in examined cells) hybridize to their complementary sequences in the spots due to the preferential binding - complementary single stranded nucleic acid sequences tend to attract to each other and the longer the complementary parts, the stronger the attraction. Such strategy allows to estimate in which sample the corresponding sequence is more abundant, however it is not suitable for measurement of the expression level in examined samples.

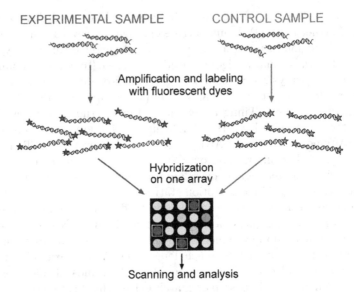

Fig. 9.4 A diagram of a typical two-color microarray experiment. Only a small fragment of the array is shown: position of spots in which more experimental (red) sample has been hybridized is marked by white squares. This means that in the experimental sample the corresponding sequence is more abundant than in control sample. When corresponding sequences are in equal amount in both samples - a yellow spot is observed.

Two-color microarrays are commercially available (e.g. "Dual-Mode" platform from Agilent, "DualChip" platform from Eppendorf, and "Arrayit" from TeleChem International). Arrays may also be easily customized for each experiment and printed on a smaller scale.

9.2.1.2   One-Color Microarrays

One-color (single-channel) microarrays, i.e. the arrays designed for hybridization with one labeled sample, provide intensity data for each probe indicating a relative level of hybridization with the labeled target. They do not truly indicate abundance levels of gene expression but rather relative abundance when compared to other samples or conditions when processed in the same experiment. Among popular single-channel systems are the Affymetrix "Gene Chip", Illumina "Bead Chip", Agilent single-channel arrays, the Applied Microarrays "CodeLink" arrays, and the Eppendorf "DualChip & Silverquant".

The procedure of sample preparation for hybridization on one-color array is essentially the same as for two-color array. For expression analysis RNA has to be reverse transcribed, amplified and labeled. After fragmentation the sample is hybridized overnight to the array, washed, stained and scanned. Obtained hybridized microarray images contain the raw data (Fig. 9.5).

To obtain information about gene expression each spot on the array should be identified, its intensity measured and normalized to make data from different arrays comparable.

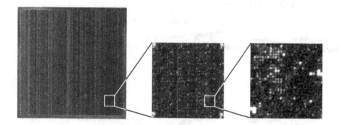

Fig. 9.5 An example of the raw data obtained from Affymetrix one-color array (in this example - from Promoter Tiling Array). From left to right larger magnifications are shown. White spots visible on the largest magnification represent individual probes which hybridized with labeled sample.

9.3 Studies of Gene Expression Regulation by Chromatin Immunoprecipitation

The protein-DNA interactions play a crucial role in many cellular processes such as DNA replication, recombination, repair, or gene transcription. The most informative are studies of interactions that take place *in vivo*, in living cells. They can be studied by chromatin immunoprecipitation (ChIP; Fig. 9.6). ChIP is a powerful method used to determine the location of DNA binding sites on the genome for a particular protein of interest at a given moment. It is possible only if a specific antibody that can recognize protein of interest is available (immunoprecipitation is the technique of precipitating a protein antigen out of solution using an antibody that specifically binds to that particular protein). Proteins that are strongly bound to DNA (like histones) can be immunoprecipitated without cell fixation. Weakly bound proteins (including transcription factors) have to be cross-linked to the DNA that they are binding. The crosslinking is often performed by applying formaldehyde to the cells (or tissue). Following crosslinking, the cells are lysed and the chromatin is physically (ultrasound) or enzymatically (nucleases) broken into small pieces (0.2-1 kb in length); the shorter chromatin fragments the better mapping of protein binding sites could be obtained. Next, the immunoprecipitation by a specific antibody is done. In parallel, a control reaction is also carried out: with unspecific antibody or without antibody. Antibody-protein-DNA complexes are extracted from solution using bacterial protein A or protein G (that specifically binds antibodies) immobilized on agarose or magnetic beads. Samples have to be washed to remove an excess of

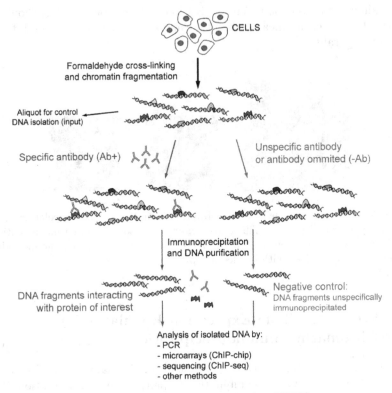

Fig. 9.6 Principles of chromatin immunoprecipitation (ChIP)

material and protein-DNA complexes are purified. Complexes are then heated
to reverse the formaldehyde cross-linking, allowing the DNA to be separated
from the proteins.

The identity and quantity of the isolated DNA fragments (both experimental and negative control) can then be determined by PCR (ch. 4.1.2).
The limitation of ChIP-PCR is that one must have an idea which genomic
region is being targeted in order to design the correct PCR primers. Instead
of PCR, a DNA microarray (ch. 9.2.1) can be used (ChIP-chip) what allows
to find on a genome-wide scale where the protein binds. The limitation of
ChIP-chip is that not all target sequences could be present on the array.
Alternatively, ChIP can be combined with sequencing (ChIP-seq). In fact,
recently emerged massive parallel sequencing (ch. 9.1.2) can localize protein
binding sites (ChIP products) in a high-throughput, cost-effective fashion.
ChIP-seq allows to find out all genomic targets of proteins of interest, also
these that are far away of the promoters. By integrating a large number of
short reads from sequencer, highly precise ($\pm$ 50 bp) binding site localization
is obtained. Obviously, these sequences have to be mapped to the genome.

9.4 Analysis of Protein Expression

It is now known that mRNA level does not always correlate with protein content. Thus, after genomics and transcriptomics, proteomics is considered the next step in the study of biological systems. Proteomics can be defined as the qualitative and quantitative comparison of proteomes under different conditions to further unravel biological processes.

An organism's genome is quite constant (although mobile elements could introduce some differences even between cells of one organism). The proteome is much more complicated than the genome, mostly because it differs from cell to cell and from state to state - distinct genes are expressed in distinct cell types and after different stimuli. What's more, many proteins are also subjected to a wide variety of chemical modifications after translation (phosphorylation, ubiquitination, methylation, acetylation, glycosylation, etc.). Some proteins undergo several modifications, often in time-dependent combinations. The human genome is estimated to encode between 20,000 and 25,000 non-redundant proteins. However, the number of unique protein species likely increase by order of magnitude due to alternative maturation of transcripts (e.g. RNA splicing), proteolysis, and post-translational modifications. It illustrates the potential complexity one has to deal with when studying protein structure and function.

Proteins are responsible for all cellular functions. Thus, characterizing of the proteome brings the most reliable information about the actual and current state of a living biological system. Classically, antibodies to particular proteins or to their modified forms are used in biochemistry and cell biology studies. They are employed in such (not described here) protein-specific techniques as Western blotting, immunohistochemistry, immunoprecipitation, ELISA, and others. More recently methods of mass spectrometry, which are not limited to specific antibody-antigen interactions, are implemented as a standard analytical tool for characterization of proteomes.

9.4.1 MALDI Mass Spectrometry

Mass spectrometry (MS) is an analytical technique that measures the mass-to-charge ratio (m/z) of charged particles (not only peptides or proteins). This allows to establish a molecular weight of the analyzed molecule. In proteomics, by using different techniques of mass spectrometry, the molecular weights of specific peptides and proteins in complex mixtures containing just picomoles of particular molecules can be read. The practical accuracy of measurement of molecular weights from mass spectra usually reaches 0.01%, while in the case of electrophoretic techniques such accuracy is somewhere between 5 and 10%. A typical mass spectrometer consists of an ion source, the analyzer, and detector, combined with computerized control system for recording and analysis of data. Depending on a type of the material under analysis (the analyte) various techniques of ionization and separation of ions in the analyzers are used. Two ionization methods: matrix-assisted laser(-induced-)desorption ionization (MALDI) and electrospray ionization (ESI) are the most commonly used for protein analyses.

MALDI spectrometers are used for the analysis of proteins and peptides with molecular weights ranging from 400 Da to as much as several hundred thousand Da. In the MALDI spectrometers the analyte molecules (e.g., protein or peptide) are co-crystallized with matrix molecules that absorb UV light (cinnamic acid derivatives are the most frequently used for protein studies). As a result of irradiation of a matrix and analyte molecules mixture by a nanosecond laser pulse, protonated molecules of the analyte are desorbed from the crystals (matrix molecules serve as a source of protons). Most of the laser energy is absorbed by the matrix what prevents unwanted fragmentation of the analyte molecules, which are usually endowed with a single protons and form $[M+H]^+$ molecular ions. Similar principle of ionization is used in Surface-Enhanced Laser Desorption Ionization (SELDI) spectrometers. In this type of spectrometers the laser beam-induced desorption of analyte molecules occurs from different types of surfaces, which specifically bind different fractions of proteins (such surfaces are coated with substances which are equivalent to chromatography deposits).

MALDI (and SELDI) spectrometers are usually coupled with Time of Flight (ToF) ion analyzers, which use strong electric field to accelerate the

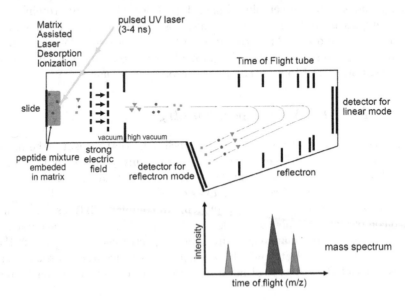

Fig. 9.7 Schematic representation of MALDI-ToF mass spectroscopy

ions. The accelerated ions are then introduced into the high vacuum flight tube and continue to fly until they reach the detector (ions should be accelerated to the same kinetic energy to ensure that all ions of identical mass move at the same speed). Light ions reach the detector earlier than heavy ions. Registered time of flight of analyzed molecular ions (typically ranging from 0.01 to 1 ms) is inversely proportional to the m/z value and can be converted into actual mass of the analytes. Resulting data are represented in a form of spectra (Fig. 9.7).

The resolution and mass accuracy are dependent on the time window wherein ions of the same mass reach the detector. This is for a large part determined by the start velocity distribution. To correct start velocity distribution the reflectron mode could be used. In this mode, the drift direction of the ions in an electric counter field is reversed (Fig. 9.7). Ions of the same mass but higher start energy drift deeper into the reflectron before being reflected and fly a longer distance before they reach the detector. In this way, they catch up with the slower moving ions at a certain point after the reflectron. The detector is located at this focusing point. The reflectron greatly enhances resolution by correcting start velocity distribution and by prolonging the flight time. Although use of the reflectron mode results in higher resolution, reflection reduces sensitivity and high molecular weight (poly)peptides can only be detected in linear mode.

The described method in its basic version allows the identification of characteristic features of the protein profiles of analyzed mixtures. However, MALDI-ToF spectrometry also allows the identification of proteins based on the amino acid sequence in polypeptide chains. Most typically, the analyzed protein is processed enzymaticaly (e.g., by trypsin digestion), then selected peptides are further fragmented in the spectrometer, and sizes (m/z values) of fragmentary ions are determined in the so called tandem mass spectrometry (MS/MS). On the basis of size of fragmentary ions the appropriate computational analysis allows the establishing of the sequence of amino acids in the peptide chain and the identification of corresponding protein. Additionally, MALDI-ToF MS can be applied directly to samples of biological tissues. While traditional methods of protein analysis require the homogenization of whole tissues, direct MS analysis of tissue requires much less sample manipulation and maintains the spatial integrity of the specimen. Thus, a single section of a tissue sample could be used not only for identification of proteins but also for determination of spatial distributions of detected molecules. Spatially-correlated MS analysis of a tissue sample that can reveal abundance of different biomolecular ions in different sample spots is called imaging mass spectrometry (IMS).

Proteomics studies generate high-dimensional spectral data sets where traditional statistical and computational methods are not sufficient for analyses, hence novel "non-standard" mathematical approaches are frequently required. A typical proteomic analysis of MS data consists of three steps: (i) preprocessing of mass spectra data, (ii) identification of spectral components,

(iii) statistical analyses and classification based on identified components. The first step includes: "smoothing" of the spectrum, removing the baseline, normalization and averaging of technical replicates. The simplest method of spectra "smoothing" is averaging among several neighboring points. The baseline correction flattens specific noise named baseline. Interpolation is performed to standardize points on the m/z axis among all spectra. Normalization is used for scaling the spectra - the most frequently used algorithms base on total ion current (TIC) value or constant noise. All these procedures reduce dimensionality of data and improve quality of further analyses. There are several algorithms accepted for these procedures and most of proteomics studies used comparable methods for spectra pre-processing. However, there is no single protocol generally accepted as the standard. In fact, some of the key parameters are tuned and adjusted for particular analyses, and type and order of pre-processing steps may significantly differ among different analyses. There are two general types of approaches allowing comparison of multiple spectra. One of them bases on comparing the signal in successive measurement points (called point-to-point analysis). The second one uses identification and matching of spectral components. This approach usually starts from peak detection. The aim of peak detection procedures is to find peaks describing the composition of spectra, which is followed by alignment of peaks identified in different spectra. Most of algorithms used for peak detection identify local maxima and minima; local maxima are classified as peaks only when they have signal to noise ratio above a given threshold. Alternative procedure for identification of spectral components bases on modeling of spectra as a sum of Gaussian components, then the model is adjusted to experimental data. Such procedure avoids artifacts linked with the methods of point-to-point and peak alignment, and facilitates quantitative and statistical analysis. The objective of the last step of spectral data analysis is to find patterns in the data sets and to classify samples. Many of the statistical and analytical algorithms and tools developed for functional genomics are being used in proteomics. Different combinations of unsupervised and supervised clustering, machine learning, pattern recognition, statistical analysis, modeling techniques, and database mining are used in the field. Some researchers use custom-build protocols and algorithms while others base on commercially available software for spectra analysis. However, all these computational tools have to be tuned and optimized for specific applications and datasets, which apparently add diversity to these protocols.

Acknowledgments. This work was inspired by Prof. Andrzej Polański and supported by the Polish National Science Centre (UMO-2011/01/B/ST6/06868) and by the European Community from the European Social Fund within the INTERKADRA project (UDA - POKL-04.01.01-00-014/10-00).

References

1. Bednár, M.: DNA microarray technology and application. Medical Science Monitor 6, 796–800 (2000)
2. Duncan, M.W., Roder, H., Hunsucker, S.W.: Quantitative matrix-assisted laser desorption/ionization mass spectrometry. Brief Funct Genomic Proteomic 7, 355–370 (2008)
3. Hacia, J.G., Collins, F.S.: Mutational analysis using oligonucleotide microarrays. Journal of Medical Genetics 36, 730–736 (1999)
4. Hoffman, B.G., Jones, S.J.M.: Identification of DNA-protein interactions. Journal of Endocrinology 201, 1–13 (2009)
5. Jamesdaniel, S., Salvi, R., Coling, D.: Auditory proteomics: methods, accomplishments and challenges. Brain Research 1277, 24–36 (2009)
6. Mardis, E.R.: Anticipating the 1,000 dollar genome. Genome Biology 7, 112 (2006)
7. Murphy, D.: Gene expression studies using microarrays: principles, problems, and prospects. Advances in Physiology Education 26, 256–270 (2002)
8. Pareek, C.S., Smoczynski, R., Tretyn, A.: Sequencing technologies and genome sequencing. Journal of Applied Genetics 52, 413–435 (2011)
9. Park, P.J.: ChIP-seq: advantages and challenges of a maturing technology. Nature Reviews Genetics 10, 669–680 (2009)
10. Ragoussis, J., Elvidge, G.P., Kaur, K., Colella, S.: Matrix-assisted laser desorption/ionisation, time-of-flight mass spectrometry in genomics research. PLoS Genetics 2, e100 (2006)
11. Roth, C.M.: Quantifying gene expression. Current Issues of Molecular Biology 4, 93–100 (2002)
12. Schadt, E.E., Turner, S., Kasarskis, A.: A window into third-generation sequencing. Human Molecular Genetics 19(R2), R227–R240 (2010)
13. Thompson, J.F., Milos, P.M.: The properties and applications of single-molecule DNA sequencing. Genome Biology 12, 217 (2011)